OCR
A LEVEL

FURTHER MATHEMATICS A

DISCRETE

Series editor
Heather Davis
Consultant Editor
Jan Dangerfield
Author
Nick Geere

HODDER
EDUCATION
AN HACHETTE UK COMPANY

Although every effort has been made to ensure that website addresses are correct at time of going to press, Hodder Education cannot be held responsible for the content of any website mentioned in this book. It is sometimes possible to find a relocated web page by typing in the address of the home page for a website in the URL window of your browser.

Orders: please contact Hachette UK Distribution, Hely Hutchinson Centre, Milton Road, Didcot, Oxfordshire, OX11 7HH. Telephone: +44 (0)1235 827827. Email education@hachette.co.uk Lines are open from 9 a.m. to 5 p.m., Monday to Friday. You can also order through our website: www.hoddereducation.co.uk

© Nick Geere 2018

Published by Hodder Education

An Hachette UK Company

Carmelite House, 50 Victoria Embankment, London EC4Y 0DZ

Impression number   5 4 3

Year   2021

Cover photo © salajean/stock.adobe.com

Illustrations by Aptara Inc.

Typeset in bemboStd 11/13 pts. by Aptara Inc.

Printed and bound by CPI Group (UK) Ltd, Croydon, CR0 4YY

A catalogue record for this title is available from the British Library

ISBN 9781510433373

Contents

Getting the most from this book

Mathematics is not only a beautiful and exciting subject in its own right but also one that underpins many other branches of learning. It is consequently fundamental to our national wellbeing.

This book covers the Discrete Mathematics elements in the OCRA AS and A Level Further Mathematics specifications. Students start these courses at a variety of stages. Some embark on AS Further Mathematics in Year 12, straight after GCSE, taking it alongside AS Mathematics, and so have no prior experience of A Level Mathematics. In contrast, others only begin Further Mathematics when they have completed the full A Level Mathematics course. This book requires no prior knowledge of A level Mathematics and so can be started at any time. Both AS and A level content is included in each chapter, with the sections on the AS content generally coming first. There is more detail on the split in each chapter in the section on Prior Knowledge.

Between 2014 and 2016 A Level Mathematics and Further Mathematics were very substantially revised, for first teaching in 2017. Changes that particularly affect Discrete Mathematics include increased emphasis on

- Problem solving

- Mathematical rigour

- Use of ICT

- Modelling.

This book embraces these ideas. A large number of exercises involve elements of problem solving and require the application of the ideas and techniques in a wide variety of real world contexts. This develops independent thinking and builds on thorough understanding. Discrete Mathematics often provides descriptions of real world situations that make them tractable to calculations, and so modelling is key to this branch of mathematics. It pervades much of the book, particularly the chapters on the use of graphs to solve real world problems.

Throughout the book the emphasis is on understanding and interpretation rather than mere routine calculations, but the various exercises do nonetheless provide plenty of scope for practising basic techniques. The exercise questions are split into three bands. Band 1 questions (indicated by a light grey) are designed to reinforce basic understanding; Band 2 questions (a darker bar) are broadly typical of what might be expected in an examination; Band 3 questions (a darker bar again) explore around the topic and some of them are rather more demanding. In addition, extensive online support, including further questions, is available by subscription to MEI's Integral website, http://integralmaths.org.

At the end of each chapter there is a list of key points covered, as well as a summary of the new knowledge (learning outcomes) that readers should have gained.

Two common features of the book are Activities and Discussion points. These serve rather different purposes. The Activities are designed to help readers get into the thought processes of the new work that they are about to meet; having done an Activity, what follows will seem much easier. The Discussion points invite readers to talk about particular points with their fellow students and their teacher and so enhance their understanding.

Answers to all exercise questions are provided at the back of the book, and also online at www.hoddereducation.co.uk/OCRFurtherMathsDiscrete.

Prior knowledge

No prior knowledge of Discrete Mathematics is needed for this book. It does, however, assume that the reader is reasonably fluent in basic algebra and graphs: working with formulae and expressions; solving linear simultaneous equations; graphing inequalities.

Matrices are used to store information but manipulation of them is not required.

Chapter 1 Solving problems

This chapter establishes a more formal treatment of a range of problem solving strategies, mainly involving counting problems. Sections 1.1, 1.2 and 1.3 cover the material for AS Further Mathematics, including arrangements and selections that are also met in A level Mathematics, Year 1, although the work can be accessed from GCSE ideas. The fourth section develops these ideas further at A level.

Chapter 2 Graphs and networks

This chapter introduces the language of graphs and networks that is developed in later chapters. The first three sections cover the AS material and the last two are required for the A level work.

Chapter 3 Algorithms

This chapter considers the characteristics of algorithms and how they work, in preparation for later work on specific examples of algorithms. Sections 3.1, 3.2, 3.3 and 3.4 cover the AS material and this is required for the A level work in section 3.5. The AS material can all be accessed from GCSE. Knowledge of the logarithmic function and factorials is required for the A level work.

Chapter 4 Network algorithms

This builds on the work in Chapter 2 but otherwise can be accessed from GCSE. Matrices are used, but only for storing information so no prior knowledge is needed. Sections 4.1, 4.2 and 4.3 contain the AS material and sections 4.4 and 4.5 contain the A level material.

Chapter 5 Critical path analysis

The work here depends only on GCSE work. Sections 5.1 and 5.2 cover the AS work and the final section on cascade charts is for A level only.

Chapter 6 Linear programming

The first section covers the AS material and requires the use of linear graphs and inequalities. The other two sections are A level work and rely on the use of ideas related to solving simultaneous equations. The idea of a matrix is useful for understanding the simplex algorithm.

Chapter 7 Game theory

This uses work on linear equations from GCSE and work on linear programming from Chapter 6. Sections 7.1 and 7.2 cover the AS material and the rest of the chapter is A level.

Solving problems

➜ Place the digits 1 to 9 in the cells of the figure below so that each row and column add to the same amount.

1 Solving problems

The ideas in this first section permeate the whole of the book. In each chapter you will be able to see how they apply in that context.

The first step is to classify problems according to their nature.

> An **existence** problem is about determining whether there is a solution.
>
> A **construction** problem involves finding a solution.
>
> An **enumeration** problem involves finding how many solutions there are.
>
> An **optimisation** problem is about finding the best solution, according to some measure, perhaps the shortest or cheapest or most profitable.

The same context can generate examples of each type of problem. The context of selecting a drink and something to eat for breakfast provides examples of the differences between these types of problems.

Suppose breakfast consists of a drink and something to eat. You can choose either tea or coffee to drink and porridge, toast or pancakes to eat.

The existence problem asks if there is a possible breakfast, to which the answer is yes, there is, because it is possible to select a drink and something to eat. No further justification is necessary.

The construction problem asks if a breakfast can be identified. Yes, by choosing one of the drinks, tea say, and one of the items to eat, porridge for example, you can identify a breakfast consisting of a drink of tea and some porridge.

The enumeration problem asks how many different breakfasts are possible. Since there are two possibilities for the drink and three for the food item, the **product rule for counting** gives a total of six different breakfasts.

The product rule for counting that you met at GCSE is also known as the **multiplicative principle**. It means that you multiply the numbers of options at each stage to work out the total number of ways something can be done.

The optimisation problem could take several forms in this context. You may wish to have the healthiest breakfast, or the one that provides the most energy, or the one that is quickest to prepare, depending on what aspect you wish to optimise.

Most of the time you would simply choose your breakfast, but there may be occasions when there are other considerations such as those described.

ACTIVITY 1.1

Figure 1.1 shows the roads connecting four towns, A, B, C and D. The lengths of the roads are shown by numbers on the lines.

Write:

(i) an existence problem
(ii) a construction problem
(iii) an enumeration problem
(iv) an optimisation problem

for the context shown.

You will meet more route problems in Chapter 4.

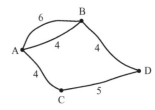

Figure 1.1

Example 1.1

For each of the following problems, decide whether it is an existence, construction, enumeration or optimisation problem, and justify your answer.

(i) How can Ian travel from Dorchester to Birmingham by train?

(ii) What is the quickest way for Ali to get to Manchester?

(iii) Can Heather fly directly from Newquay to Norwich?

Note

Many problems will be in more than one category. For example, an optimisation problem is likely also to involve constructing the solution. An existence problem may be solved by constructing a solution.

Note

The pigeonhole principle can be stated in other ways, such as: If m pigeons are placed in m pigeonholes then there is an empty hole if, and only if, there is a hole with more than one pigeon.

Solution

(i) This is a construction problem as Ian is seeking a route by rail between Dorchester and Birmingham.

(ii) This is an optimisation problem as Ali wants the fastest route to Manchester.

(iii) This is an existence problem as Heather wishes to know whether such a route exists.

Existence problems are explored further below and in Chapters 2, 3, 4, 6 and 7.

Construction problems are explored further in Chapters 2, 3, 4, 5, 6 and 7.

Enumeration problems are explored further in Sections 1.3 and 1.4 and Chapter 3.

Optimisation problems are explored further in Chapters 3, 4, 5, 6 and 7.

The pigeonhole principle

Some problems can be solved using the **pigeonhole principle**, which states:

If there are m pigeonholes and n pigeons, where $m < n$, then there will be at least one pigeonhole with more than one pigeon.

This seems an obvious idea but it is a powerful technique for solving problems.

You may have met the sock problem before.

A drawer contains red socks and blue socks. They are identical except for their colour. How many socks must you draw, at random, to be certain of a matching pair?

The answer is three socks. The 'pigeonholes' are 'blue' and 'red' and the 'pigeons' are the socks. Since there are two colours (pigeonholes), and three socks (pigeons), and 3 > 2 there must be more than one sock in at least one of the colours.

The key to using the pigeonhole principle to solve a problem is identifying the 'pigeonholes' and the 'pigeons'. This can be quite difficult! It requires creative thinking.

ACTIVITY 1.2
How many socks would you need to draw if there were five different colours of sock?

Example 1.2

On every square of a 7 by 7 board there is a flea. The fleas simultaneously all jump to an adjacent square (that is, a square with an edge in common with the original one). Is there now one flea on every square?

Note

This is an existence problem.

Solution

Colour the board black and white as for a chess board, with the corners being white squares. The fleas on white squares will jump onto black squares. The fleas on black squares will jump onto white squares.

The number of black squares is one less than the number of white squares. By the pigeonhole principle there will be at least one white square with no flea after they have jumped. Similarly, there will be at least one black square with more than one flea after they have jumped.

So, there will not be one flea on every square after they have jumped.

Example 1.3

Seven numbers are selected from the numbers 1 to 12. Show that at least two of them add to 13.

Solution

Pair the numbers from 1 to 12 so that each pair adds to 13.

Label some imaginary boxes using those pairs.

These are the pigeonholes. →

| 1, 12 | 2, 11 | 3, 10 | 4, 9 | 5, 8 | 6, 7 |

The seven numbers are the pigeons. →

Place each of the seven numbers into the box that is labelled with that number.

Since there are seven numbers and six boxes, at least one of the boxes will contain two numbers and hence two of the seven numbers must add to 13.

Exercise 1.1

① Sehrish is researching to see if it is possible to get from Newcastle to London in less than 2 hours.

   (i)   What type of problem is this?

   (ii)   Rewrite this as an optimisation problem.

② Show that for any group of 11 PIN numbers (a PIN number consists of four digits, that may be repeated), two of the PIN numbers must end with the same digit.

③ Figure 1.2 shows a mystic rose with seven points. Each point is joined by a straight line to every other point.

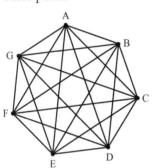

Figure 1.2

   (i)   Write an existence problem based on Figure 1.2.

   (ii)   Write a construction problem based on Figure 1.2.

   (iii)   Write an enumeration problem based on Figure 1.2.

   (iv)   Write an optimisation problem based on Figure 1.2.

④ Show that for any five integers selected from the numbers 1 to 8, there is a pair with an odd sum.

⑤ An extension of the pigeonhole principle states that: 'For any non-empty, finite set of real numbers, the maximum value is at least the mean value.' (Dijkstra)

Hence show that, if the integers 1 to 10 are written in a circle, there are three adjacent numbers whose sum is ≥ 17.

2 Set theory

Set theory can be used to represent problems.

Figure 1.3 shows a Venn diagram.

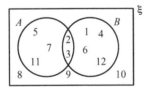

Figure 1.3

The **universal set** in this case, $\xi = \{1, 2, 3, 4, 5, 6, 7, 8, 9, 10, 11, 12\}$ and contains all of the **elements**, in this case integers, that are being considered.

Set A $= \{2, 3, 5, 7, 11\}$ or the set of prime numbers, $3 \in$ A (3 is a member of set A) and $4 \notin$ A (4 is not a member of set A).

$A' = \{1, 4, 6, 8, 9, 10, 12\}$ is the **complement** of A and consists of the elements that are not in set A.

$n(A) = 5$, the number of elements in set A.

The **empty** set, or **null** set, can be written as { } or ϕ and contains no elements, so $n(\phi) = 0$.

ACTIVITY 1.3

(i) Write down
    (a) the elements of set B
    (b) B′
    (c) n(B).

(ii) How could you describe the elements of set B?

> A proper subset is part of a set that is smaller than that set.

$\{1, 2\} \subseteq B$ means that $\{1, 2\}$ is a **subset** of B, $\{1, 2\} \subset B$ means that $\{1, 2\}$ is a **proper subset** of B.

So, $B \subseteq B$ but $B \not\subset B$.

$\{2, 3\} = A \cap B$ and is the **intersection** of A and B.

$\{1, 2, 3, 4, 5, 6, 7, 11, 12\} = A \cup B$ and is the **union** of A and B.

Example 1.4

(i) Write down $n(A \cap B)$ and $n(A \cup B)$.

(ii) Verify that $n(A \cup B) = n(A) + n(B) - n(A \cap B)$.

> **Discussion point**
> → Is the inclusion-exclusion principle true for any pair of sets?

Solution

(i) $(A \cap B) = 2, n(A \cup B) = 9$

(ii) $n(A) + n(B) - n(A \cap B) = 5 + 6 - 2 = 9 = n(A \cup B)$ so the result is true.

$n(A \cup B) = n(A) + n(B) - n(A \cap B)$ is the **inclusion-exclusion principle**.

It can be used to solve problems.

Example 1.5

A group of 60 people in a Book Club have either read 'War and Peace' or 'Les Miserables' or both. 42 people have read 'Les Miserables' and 31 people have read 'War and Peace'. How many have read both?

Solution

Let A = set of those who have read 'Les Miserables', and B = set of those who have read 'War and Peace'.

By the inclusion-exclusion principle,

$n(A \cup B) = n(A) + n(B) - n(A \cap B)$

$n(A \cup B) = 60$ ◄── A ∪ B contains all of the people.

$n(A) = 42$ ◄── A contains those who have read 'Les Miserables'.

$n(B) = 31$ ◄── B contains the people who have read 'War and Peace'.

$n(A \cap B)$ represents the people who have read both.

$60 = 42 + 31 - n(A \cap B) \Rightarrow n(A \cap B) = 13$

So 13 people have read both.

The people in Example 1.5 could be separated into 3 subsets:

- People who have read only 'War and Peace'.
- People who have read only 'Les Miserables'.
- People who have read both.

These sets are mutually exclusive and exhaustive so each one of the original 60 people is in exactly one of the subsets.

The three subsets, together, represent a **partition** of the original set:

{people who have read only 'War and Peace' | people who have read only 'Les Miserables' | people who have read both books}.

Example 1.6

Write down all the partitions of $\{1, 2, 3\}$.

Note

Finding how many partitions there are of a set is an enumeration problem.

Solution

$\{1 \mid 2 \mid 3\}, \{1 \mid 2, 3\}, \{2 \mid 1, 3\}, \{3 \mid 1, 2\}, \{1, 2, 3\}$

There are five partitions altogether.

Note

The order of the subsets does not matter, so e.g. {1|2, 3} is the same as {2, 3|1}, {1|3, 2} and {3, 2|1}.

ACTIVITY 1.4

How many partitions are there of the set {1, 2, 3, 4, 5}?

The problem of finding all of the possible partitions of a set is a substantial enumeration problem. Section 1.3, on arrangements and selections, deals with ways of counting that do not require listing all the possibilities.

You should be familiar with the following sets.

N	the set of natural numbers $\{1, 2, 3, \ldots\}$
$\mathbb{Z}$	the set of integers $\{\ldots -2, -1, 0, 1, 2, 3, \ldots\}$
$\mathbb{Q}$	the set of rational numbers (that can be written in the form $\frac{a}{b}$, where a and b are integers)
$\mathbb{R}$	the set of real numbers (any number along the x axis, for example)
$\mathbb{C}$	the set of complex numbers (any number with a real and imaginary part)
$\mathbb{Z}^+$	the set of positive integers $\{1, 2, 3, \ldots\}$
$\mathbb{Z}_0^+$	the set of non-negative integers $\{0, 1, 2, 3, \ldots\}$

Table 1.1

$\{\mathbb{Z}^+ \mid \{0\} \mid \mathbb{Z}^-\}$ forms a partition of $\mathbb{Z}$.

① For each of the Venn diagrams, select the descriptions that apply to it.

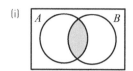
(i)

Figure 1.4

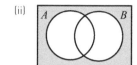
(ii)

Figure 1.5

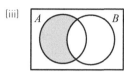
(iii)

Figure 1.6

(a) $(A \cup B)'$ (b) $A \cap B$ (c) $(A \cup B) \cap B'$ (d) $(A' \cup B')'$ (e) $A' \cap B'$ (f) $A \cap B'$

② How many numbers between 1 and 50 are not divisible by 2 or 5?

③ Is $\{\mathbb{Z} \mid \mathbb{Q}\}$ a partition of $\mathbb{R}$? Justify your answer.

④ A group of 80 members of a sports club are asked which sports they like. There are 35 who like tennis and 64 who like badminton and all 80 like at least one of the sports. How many like both sports?

⑤ Two sets contain 23 and 34 items respectively. The intersection of the two sets contains 19 items. How many items are contained in the union of the two sets?

3 Arrangements and selections

This section is concerned with enumeration problems, counting the number of ways of arranging items or selecting groups of items.

A password, consisting of the four letters A, B, C, D is needed for a lock. These letters can be ordered in several different ways: for example, ACDB, CADB and so on, but the letters may not be repeated. To determine how many different ways there are, consider each position as a place holder.

> The first place can be filled in four different ways as there are four letters.

| 4 | 3 | 2 | 1 |

> There are then three letters left to choose from to fill the next place.

> There are two ways of filling the third place.

> There is only one way for the final place as there is only one letter remaining.

So $4 \times 3 \times 2 \times 1 = 24$ is the number of ways of arranging the four letters.

$4 \times 3 \times 2 \times 1$ is denoted by 4!, and is called 4 **factorial**.

In general, $n! = n \times (n-1) \times (n-2) \ldots \times 3 \times 2 \times 1$

Note

This uses the multiplicative principle, also known as the product rule for counting.

ACTIVITY 1.5
Verify that there are 24 different arrangements of A, B, C and D.

Only 24 different arrangements does not make for a secure lock. A better way would be to choose from a greater number of letters. Suppose the letters can be chosen from A, B, C, D, E, F, G, H, I, J. The lock still requires each letter to be different.

| The first place can be filled in ten different ways. | → | 10 | 9 | 8 | 7 |

There are then nine different ways of filling the next place.

There are eight ways for the third place.

There are seven ways of filling the final place.

So $10 \times 9 \times 8 \times 7 = 5040$ is the number of ways of arranging the four letters, giving a more secure code.

Now, $10 \times 9 \times 8 \times 7 = \dfrac{10 \times 9 \times 8 \times 7 \times 6 \times 5 \times 4 \times 3 \times 2 \times 1}{6 \times 5 \times 4 \times 3 \times 2 \times 1} = \dfrac{10!}{6!} = \dfrac{10!}{(10-4)!}$

In general, the numbers of ways of arranging r objects, chosen from n objects is:

$n \times (n-1) \times (n-2) \times \ldots \times (n-r+1) = \dfrac{n!}{(n-r)!}$ and is denoted ${}_nP_r$ or ${}^nP_r$.

It is the number of **permutations** of r objects selected from n objects.

In this case the number of different codes is ${}^{10}P_4$.

Note

Find out how your calculator works this out.

○**ACTIVITY 1.6**
Find the number of different combinations if the letters A to J may be repeated in the code.

Example 1.7

Find the number of distinct arrangements of the letters of the word BOOKS.

Solution

There are five letters in the word and so there are 5! ways of arranging them.

However, two of the letters are the same and so some of those 5! ways will be indistinguishable from each other. ← | OO looks the same as OO. |

There are two ways of arranging the two Os and so every arrangement of the letters appears twice.

The number of distinct arrangements = 5! ÷ 2 = 60.

In general, for each group of k items that are indistinguishable from each other, divide the total number of arrangements by $k!$.

Example 1.8

How many distinct arrangements are there of the letters in the word MATHEMATICAL?

Solution

There are twelve letters, including two Ms, two Ts and three As.

Total number = 12! ÷ (2!) ÷ (2!) ÷ (3!) = 19 958 400.

Example 1.9

Charlie has six mathematics books and four books on gardening. He wants to keep the same type of books together on his shelf. In how many ways can he arrange them?

Solution

The six mathematics books can be arranged in 6! ways.

The four gardening books can be arranged in 4! ways.

He can either have the mathematics books first, or the gardening books.

So, the total number of arrangements = $6! \times 4! \times 2 = 34560$.

Selections

The problems that have been considered so far concern situations where the order of the items matters. When you choose some students to work in a group together the order is unimportant. Suppose you require five students and you choose from a class of twelve.

The number of arrangements = $^{12}P_5 = \frac{12!}{7!}$.

However, for each group of five students there are 5! ways in which they can be arranged. The number of arrangements is 5! times too large.

The number of groups = $\frac{12!}{7!} \div 5! = \frac{12!}{7!5!}$.

This may be written as $^{12}C_5$.

In general, the number of **combinations**, or selections, of r objects, chosen from n objects is:

$$^nC_r = \frac{n!}{(n-r)!r!}$$

Example 1.10

A football team consisting of one goalkeeper, four defenders, three midfielders and three forwards is to be selected from a squad of two goalkeepers, five defenders, five midfielders and six forwards. How many different teams can be chosen?

Solution

The goalkeeper can be selected in $^2C_1$ ways.

The defenders can be selected in $^5C_4$ ways.

The midfielders can be selected in $^5C_3$ ways.

The forwards can be selected in $^6C_3$ ways.

The total number of teams = $^2C_1 \times {}^5C_4 \times {}^5C_3 \times {}^6C_3$

$$= 2 \times 5 \times 10 \times 20$$

$$= 2000$$

Exercise 1.3

① Find the number of different arrangements of the letters of the word NUMERICAL.

② There are eight horses in a race. How many ways are there for the horses to fill the first three places in the race?

③ How many different six-digit numbers can be made using the digits of 233123?

④ Find how many ways a team of four can be chosen from a group of nine players.

⑤ In how many ways can three boys and four girls seat themselves in a straight line if:

   (i)   the boys all sit together and the girls all sit together?

   (ii)   boys and girls seat themselves alternately?

⑥ Four friends go out for a meal together. The table is arranged so that two can be seated on one side and two opposite them. Two of the friends don't want to sit opposite each other. How many ways can the friends be seated?

⑦ How many five-digit numbers greater than 30 000 can be made from the digits 0, 1, 2, 3 and 4 if no digits may be used twice?

⑧ How many 'words' of three of the letters of the word PHONES are there that contain at least one vowel?

⑨ A bracelet is made of ten coloured beads. All of the beads are different colours.

   How many different bracelets are possible if:

   (i)   there is no restriction on the design?

   (ii)   the red bead and the pink bead may not be next to each other?

⑩ A rugby team of 15 people is to be selected from a squad of 25 players.

   (i)   How many different teams are possible?

   In fact, the team has to consist of 8 forwards and 7 backs.

   (ii)   If 13 of the squad are forwards and the other 12 are backs, how many different teams are now possible?   [MEI]

4 Further problems

Venn diagrams with three sets

The inclusion-exclusion principle extends to three sets.

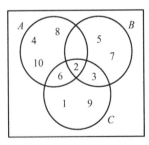

Figure 1.7

A = {$x \in \mathbb{N}$, $x \leqslant 10$ and x is even}.

B = {$x \in \mathbb{N}$, $x \leqslant 10$ and x is prime}.

C = {$x \in \mathbb{N}$, $x \leqslant 10$ and x is a factor of 18}.

$n(A \cup B \cup C) = n(A) + n(B) + n(C) - n(A \cap B) - n(B \cap C) - n(C \cap A) + n(A \cap B \cap C)$

ACTIVITY 1.7
Verify that the inclusion-exclusion principle works for the Venn diagram in Figure 1.7.

Example 1.11

50 students are asked what flavour of ice cream they like. 20 like vanilla, 17 like strawberry and 33 like chocolate. 12 like both vanilla and chocolate, 11 like strawberry and vanilla and 7 like strawberry and chocolate. How many like all three flavours?

Solution

Let V = {those who like vanilla ice cream}, S = {those who like strawberry ice cream} and C = {those who like chocolate ice cream}.

$n(V \cup S \cup C) = n(V) + n(S) + n(C) - n(V \cap S) - n(S \cap C) - n(C \cap V) + n(V \cap S \cap C)$

Substituting gives:

$50 = 20 + 17 + 33 - 11 - 7 - 12 + n(V \cap S \cap C)$

So, the number who like all three is 10.

Example 1.12

A group of eight people are seated around a circular table. There are four men and four women. How many arrangements are there in which the men and women alternate?

Solution

The four men can be arranged in 4! ways and so can the four women. Since they have to alternate there are 4! × 4! ways of arranging them if the first person is, say, a man and they are in a straight line. They are, however, in a circle, and so there are four positions that appear identical in terms of the arrangement.

Number of ways = 4! × 4! ÷ 4 = 144

Derangements

A **derangement** is a permutation such that no object is in its original position.

Example 1.13

In how many ways can 1234 be rearranged so that no digit is in its original position?

 Note

It can be shown that the number of derangements is approximately 37% of the number of arrangements, rounded to the nearest integer. This is because $0.37 \approx \frac{1}{e}$. You may wish to research further!

Solution

Clearly 1 cannot be in the first position.

Suppose 2 is in the first position:

2 _ _ _ , place 1 next $\Rightarrow$ 2 1 _ _ $\Rightarrow$ 2143 as 3 may not be in the third position.

2 _ _ _ , place 3 next $\Rightarrow$ 2 3 _ _ $\Rightarrow$ 2341 as 4 may not be in the fourth position.

→

$2\,\_\,\_\,\_$, place 4 next $\Rightarrow 2\,4\,\_\,\_ \Rightarrow 2413$ as 3 may not be in the third position.

Suppose 3 is in the first position:

$3\,\_\,\_\,\_$, place 1 next $\Rightarrow 3\,1\,\_\,\_ \Rightarrow 3142$ as 4 may not be in the fourth position.

2 cannot be placed in the second position.

$3\,\_\,\_\,\_$, place 4 next $\Rightarrow 3\,4\,\_\,\_ \Rightarrow 3412$

$3\,\_\,\_\,\_$, place 4 next $\Rightarrow 3\,4\,\_\,\_ \Rightarrow 3421$

Suppose 4 is in the first position:

$4\,\_\,\_\,\_$, place 1 next $\Rightarrow 4\,1\,\_\,\_ \Rightarrow 4123$ as 3 may not be in the third position.

2 cannot be placed in the second position.

$4\,\_\,\_\,\_$, place 3 next $\Rightarrow 4\,3\,\_\,\_ \Rightarrow 4312$

$4\,\_\,\_\,\_$, place 3 next $\Rightarrow 4\,3\,\_\,\_ \Rightarrow 4321$

This gives 9 derangements of the digits 1234 and so 9 arrangements where no number is in its original position.

Exercise 1.4

① How many integers between 1 and 100, inclusive, are not divisible by 2, 3 or 5?

② How many even five-digit numbers greater than 30 000 can be made from the digits 0, 1, 2, 3, 4 and 5 if no digits may be used twice?

③ In how many ways can the letters of the word SMILE be arranged so that no letter is in its original position?

④ A code is formed of three different digits, from 0 to 9, in ascending order. How many different codes are there?

⑤ Eighty students are asked what subjects they are studying. Forty study French, 35 study History and 50 study Physics. None of the students studies all three subjects and 32 study French and History. How many students study Physics and one of the other two subjects?

⑥ A group of six friends wish to save money on gifts at Christmas. They plan a 'Secret Santa' where each person buys a gift for one other person in the group, but who gives the gift must remain secret. How many ways are there of doing this so that nobody buys a gift for themselves?

⑦ A quiz team of five students is chosen from ten girls and eight boys. The team must consist of more girls than boys. How many different teams are possible?

⑧ Five couples are dancing at a barn dance and each couple consists of a man and a woman. Half way through the dance it is necessary for them all to change partners so that everyone is dancing with a different partner, again of the opposite gender. In how many ways can this be done?

⑨ How many ways can you arrange the letters of the word SEVEN so that no letter is in its original place?

⑩ Jan has five books by Thomas Hardy, two by Charles Dickens and four by Jane Austen. She arranges them on a bookshelf so that all the books by the same author are together.

(i) In how many ways can this be done?

Sue rearranges the books so that no two books by the same author are together.

(ii) In how many ways can this be done?

LEARNING OUTCOMES

Now you have finished this chapter, you should be able to

➤ understand and use the terms 'existence', 'construction', 'enumeration' and 'optimisation' in the context of problem solving

➤ use the pigeonhole principle in solving problems

➤ understand and use the basic language and notation of sets

➤ use the inclusion-exclusion principle for two sets in solving problems

➤ understand and use the multiplicative principle

➤ recall that the number of arrangements of n distinct objects is $n!$

➤ enumerate the number of ways of obtaining an ordered subset (permutation) of r elements from a set of n distinct elements

➤ enumerate the number of ways of obtaining an unordered subset (combination) of r elements from a set of n distinct elements

➤ solve problems about enumerating the number of arrangements of objects in a line, including those involving: repetition, restriction

➤ solve problems about selections, including with constraints

➤ solve problems about enumerating the number of arrangements of only some of a group of objects

➤ solve problems with several constraints

➤ extend the inclusion-exclusion principle to more than two sets

➤ find derangements.

KEY POINTS

1 The pigeonhole principle states that: if there are m pigeonholes and n pigeons, where $m < n$ then there will be at least one pigeonhole with more than one pigeon.

2 The inclusion-exclusion principle, for two sets, states that
$n(A \cup B) = n(A) + n(B) - n(A \cap B)$.

3 A partition of a set is a way of dividing a set into subsets such that every element is in exactly one of the subsets.

4 Some commonly used sets are:

N	the set of natural numbers, $\{1, 2, 3, \dots\}$
Z	the set of integers, $\{\dots -2, -1, 0, 1, 2, 3, \dots\}$
Q	the set of rational numbers (that can be written in the form $\frac{a}{b}$, where a and b are integers)
R	the set of real numbers (any number along the x-axis, for example)
C	the set of complex numbers (any number with a real and imaginary part)

5 The number of arrangements of n objects in a line is $n!$.

6 The number of ways r objects, chosen from n objects, can be arranged in a line is $^nP_r = \dfrac{n!}{(n-r)!}$.

7 The number of groups of r objects that can be chosen from n objects is $^nC_r = \dfrac{n!}{(n-r)!r!}$.

8 The inclusion-exclusion principle for three sets states:
$n(A \cup B \cup C) = n(A) + n(B) + n(C) - n(A \cap B) - n(B \cap C) - n(C \cap A)$
$+ n(A \cap B \cap C)$.

9 A derangement is a permutation where every object is no longer in its original position.

2 Graphs and networks

→ Can you draw this diagram without lifting your pen from the paper or repeating any line?

1 The language of graphs and networks

This chapter introduces the main terminology that is used in graph and network theory. It will be needed in subsequent chapters, where some of the applications of the theory are explored.

The types of graph that are considered here are different from the graphs of functions that you will be familiar with. For example, the graph in Figure 2.1 shows connections between pairs of members of a discrete set (of towns and cities in south-west England). The connections could, for example, represent the existence of a direct bus route.

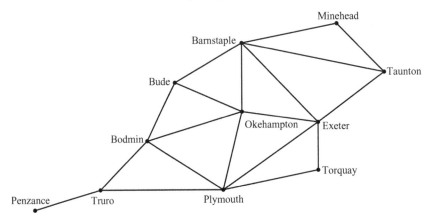

Figure 2.1

In Figure 2.1, the towns (and cities) are referred to as **vertices** (or **nodes**), whilst the connections between them are described as **edges** (or **arcs**). An edge must have a vertex at each end.

A **directed graph** (or **digraph**) is a graph where at least one edge has a direction associated with it.

A **network** is a graph with numbers associated with the edges, called weights (e.g. distances, travel times, costs).

Another example of a graph is shown in Figure 2.2. It represents the relationship 'share a common factor other than 1'.

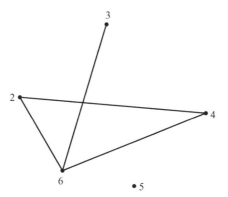

Figure 2.2

■ Within a particular graph, a **walk** is a sequence of edges in which the end of one edge is the start of the next (except for the last edge).

■ A **trail** is a walk in which no edge is repeated (but you are allowed to pass through a particular vertex more than once).

■ A trail that starts and ends at the same vertex is a **closed trail**.

■ A **path** is a trail with the further restriction that no vertex is repeated.

■ A closed path is called a **cycle**.

■ A **route** is a general term referring to a walk, trail or path, and it may be closed.

Example 2.1

Look at Figure 2.1. What description would you give to the following routes?

(i) Okehampton–Exeter–Barnstaple–Okehampton–Plymouth–Bodmin–Okehampton

(ii) Okehampton–Bodmin–Truro–Plymouth–Torquay

Solution

(i) It is at least a walk, as each edge follows the previous one.
It is at least a trail, as no edge is repeated.
It is a closed trail, as it returns to its starting point at Okehampton.
It is not a path, as the vertex Okehampton is repeated.
So it is a closed trail.

(ii) It is at least a walk, as each edge follows the previous one.
It is at least a trail, as no edge is repeated.
It is not closed, as it does not return to its starting point at Okehampton.
It is at least a path, as no vertex is repeated.
So it is a path.

■ A graph is said to be **connected** if there exists a path between every pair of vertices; i.e. if no vertices are isolated. This means that the graph in Figure 2.2 is not connected.

- It is possible for two vertices to be connected by **multiple edges**, or for a vertex to be connected to itself (forming a **loop**). These situations are shown in Figure 2.3.

Figure 2.3

- A graph that has no multiple edges or loops is referred to as a **simple graph**.

- A **tree** is a simply connected graph with no cycles. Figure 2.4 shows an example.

Discussion point
→ What is the smallest number of edges that a simply connected graph with *n* vertices can have?

Figure 2.4

- A **subgraph** is a graph that is formed from some of the vertices and edges of another graph. Note that whilst this may result in an isolated vertex, any edge has to have a vertex at each end.

- A subgraph *H*, of a connected graph *G*, is said to be a **spanning tree** of *G* if *H* is a tree and it contains all the vertices of *G*. One particular spanning tree for the graph in Figure 2.1 is shown in Figure 2.5. In general, there may be many possible spanning trees.

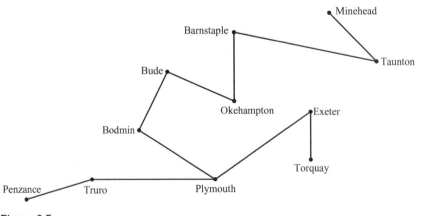

Figure 2.5

This concept will be of use in Chapter 4, where you will consider the minimum connector problem.

- The **degree** (or sometimes **order**) of a vertex is the number of edges that join it (or are **incident** to it). (A loop contributes two to the degree of its vertex.) A vertex that has an odd degree, for example, can be referred to as an odd vertex.
- In the case of a digraph, the **indegree** of a vertex is the number of edges leading to the vertex. The **outdegree** of a vertex is the number of edges leading away from the vertex.
- Two vertices are said to be **adjacent** if they are joined by an edge. Two edges are said to be adjacent if they share a common vertex.

ACTIVITY 2.1
Prove that, in a graph, the number of odd vertices is always even.

Exercise 2.1

① (i) List all the cycles in the graph below that can start and finish at A. Note that, for example, ABCA and ACBA represent the same cycle.

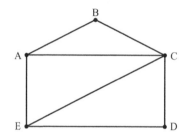

Figure 2.6

(ii) Why is ABCEDCA not a cycle?

(iii) What could ABCEDCA best be described as?

② Draw three different trees, each containing five vertices and four edges. [MEI]

③ Vertices of the graph shown in Figure 2.7 represent objects. Some edges have been drawn to connect vertices representing objects which are the same colour.

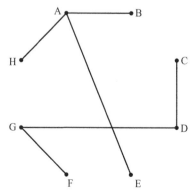

Figure 2.7

(i) Copy the diagram and draw in whichever edges you can be sure should be added.

(ii) How many edges would be needed in total if you were also told that the objects represented by B and F were the same colour? [MEI]

④ Draw a graph that is not connected and not simple, and has six vertices, all of degree 3.

⑤ Table 2.1 shows the numbers of vertices of degrees 1, 2, 3 and 4 in four different graphs. Draw an example of each of these graphs.

Degree of vertex	1	2	3	4
Graph 1	4	0	0	1
Graph 2	0	0	4	1
Graph 3	0	1	0	1
Graph 4	2	0	0	1

Table 2.1 [MEI]

⑥ A particular simply connected graph has five vertices and seven edges, and the degree of each vertex is either 2, 3 or 4.

(i) Explain why the sum of the degrees of the vertices is 14.

(ii) Copy and complete Table 2.2 to show two of the possibilities for the numbers of vertices of each degree.

Number of vertices	Number of degree 2	Number of degree 3	Number of degree 4	Sum of degrees
5				14
5				14

Table 2.2

(iii) Draw a diagram for each of your possibilities from part (ii). [MEI]

⑦ (i) A simply connected graph has seven vertices, all having the same degree d. Give the possible values of d, and for each value of d give the number of edges of the graph.

(ii) Another simply connected graph has eight vertices, all having the same degree d.

Draw such a graph with $d = 3$, and give the other possible values of d.

(iii) Explain why there are no odd values for d in (i) and why it is possible for d to be odd in (ii).

2 Types of graphs

Eulerian graphs

An interesting problem is whether it is possible to travel round a graph without repeating any edges (in other words, along a trail), so that all the edges in the graph are covered.

> If such a trail ends at its starting point, it is called **Eulerian**. If it ends somewhere else, it is called **semi-Eulerian**.

A graph that possesses an Eulerian or semi-Eulerian trail is called an Eulerian or semi-Eulerian graph, as appropriate.

A practical example of this would be a gritting truck that needs to travel down all the roads in a particular area, without repeating any roads if it can. (It is assumed that the roads are narrow, so the truck need travel in one direction only.) In the case of an Eulerian graph, the truck would be able to return to its depot, whereas in the case of a semi-Eulerian graph it would not. This is an example of the 'route inspection problem' that you will meet in Chapter 4.

> **Historical note**
>
> Leonhard Euler (1707–1783) was a very distinguished and versatile Swiss mathematician. His name is pronounced 'oiler'.

> If a graph has no odd vertices, then it can be shown to be Eulerian.

> You saw earlier in Activity 2.1 that the number of odd vertices of a graph is always even.

For every edge leading into a vertex, there will be another edge leading out, and it will be possible to move round the graph, covering each edge exactly once, without getting stranded at any vertex.

> If just two of the vertices are odd, then the graph can be shown to be semi-Eulerian.

 Note

For the start vertex, the number of outgoing edges is one greater than the number of incoming edges (and the other way round for the end vertex).

You can start at one of these odd vertices, cover all of the edges exactly once, and end up at the other odd vertex.

Further graph and network theory

Computers are often used to tackle problems involving graphs. The diagram representing a graph is not very convenient for a computer, but it is possible to completely define the essential features of a graph in a way that is usable by a computer.

An **adjacency matrix** (also known as an incidence matrix) shows the number of edges connecting any two vertices. Table 2.3 shows the adjacency matrix for the graph in Figure 2.8.

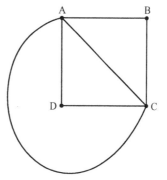

	A	B	C	D
A	0	1	2	1
B	1	0	1	0
C	2	1	0	1
D	1	0	1	0

Table 2.3

Figure 2.8

The 2 in row C and column A indicates that there are 2 edges leading from vertex C to vertex A. Notice the symmetry of the matrix about the leading diagonal (from top left to bottom right). This symmetry only applies to a non-directed graph.

Example 2.2

The adjacency matrix for a graph is shown in Table 2.4. Draw the graph it represents.

	A	B	C
A	2	0	1
B	0	2	2
C	1	2	0

Table 2.4

Solution

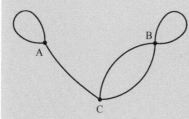

Figure 2.9

The 2 in row A and column A (for example) indicates that there are 2 edges leading from vertex A to vertex A. This counts as a single loop, as it is possible to travel either way round the loop.

A **complete** graph is one where every two vertices share exactly one edge (and where there are no loops). A complete graph with n vertices is denoted by K_n. One representation of the K_4 graph is shown in Figure 2.10.

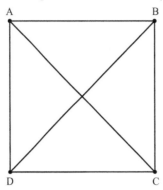

Figure 2.10

How many edges does K_n have?

Solution

Labelling the vertices 1 to n, there are $n - 1$ vertices joined by an edge to vertex 1. Excluding vertex 1, there are $n - 2$ vertices joined by an edge to vertex 2, and so on until you reach vertex $n - 1$, which has 1 edge joined to the remaining vertex.

So the total number of edges is $(n - 1) + (n - 2) + \cdots + 1 = \frac{1}{2}(n - 1)n$.

Figure 2.11 shows an example of a **bipartite** graph. The special feature of this type of graph is the division of the vertices into two sets, with edges only joining a vertex in one set to a vertex in the other.

A typical application of a bipartite graph is in allocating tasks to workers. Each edge might indicate a task that a particular worker has been trained to perform.

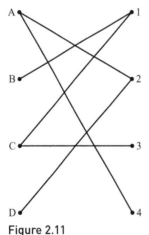

Figure 2.11

In a **complete bipartite graph** (denoted by $K_{m,n}$), each of the m vertices on one side is connected exactly once to each of the n vertices on the other.

Figure 2.12 shows the complete bipartite graph $K_{4,4}$.

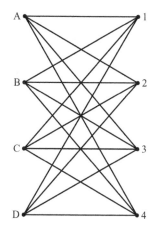

Figure 2.12

A graph is said to be **k-colourable** if each of its vertices can be assigned one of k colours in such a way that no two adjacent vertices have the same colour.

Exercise 2.2

① Determine whether the graphs below are

(i) Eulerian

(ii) semi-Eulerian

(iii) neither.

Graph 1 **Graph 2**

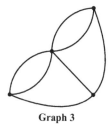

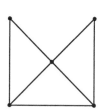

Graph 3 **Graph 4**

Figure 2.13

② Draw the graph represented by the adjacency matrix in Table 2.5.

	A	**B**	**C**	**D**
A	0	1	2	0
B	1	0	0	1
C	2	0	2	1
D	0	1	1	0

Table 2.5

③ Create an adjacency matrix for the graph in Figure 2.14.

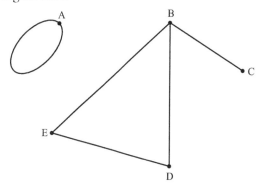

Figure 2.14

④ In the 18th century, the inhabitants of Königsberg (now Kaliningrad) enjoyed promenading across the town's seven bridges – shown in the diagram below. It was known not to be possible to cross each bridge once and once only.

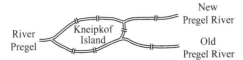

Figure 2.15

(i) Create a graph to model this situation.

(ii) How is it possible to tell that the bridges could not be crossed once and once only?

⑤ The diagram below shows a printed circuit board with two points for external connections and three internal points. Each of the connection points is to be wired to each of the internal points.

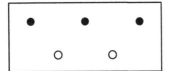

Key

● Internal point

○ Connection point

Figure 2.16

(i) Show that the two connection points can each be wired directly to each of the three internal points without any wires crossing.

(ii) Show that two connection points can be wired to four internal points without any wires crossing.

(iii) Give the smallest numbers of connection points and internal points for which at least one crossing will be required.

[MEI]

⑥ Create a matrix of weights to represent the network in Figure 2.17.

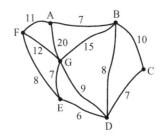

Figure 2.17

⑦ Draw the network with the matrix of weights in Table 2.6.

	A	B	C	D	E
A	–	11	9	–	12
B	11	–	14	11	8
C	9	14	–	10	–
D	–	11	10	–	7
E	12	8	–	7	–

Table 2.6

⑧ (i) When is a complete graph Eulerian or semi-Eulerian?

(ii) When is a complete bipartite graph Eulerian or semi-Eulerian?

⑨ Can you find another proof of the fact that K_n has $\frac{1}{2}(n-1)n$ edges?

⑩ (i) A, B, C and D are the vertices of the complete graph, K_4. List all the paths from A to B.

(ii) Show that there are 16 paths from A to B in the complete graph on the vertices {A, B, C, D, E}.

3 Isomorphisms

Two graphs are said to be **isomorphic** if one can be distorted in some way to produce the other (from the Greek: same form). Isomorphic graphs must have the same number of vertices, each of the same degree, and their vertices must be connected in the same way.

Consider the graphs in Figures 2.18 and 2.19. The vertices of Figure 2.18 cannot be put into a one-to-one correspondence with those of Figure 2.19 (for example, there is no vertex of degree 4 in Figure 2.18 that could correspond with vertex D of Figure 2.19). Therefore these graphs are not isomorphic.

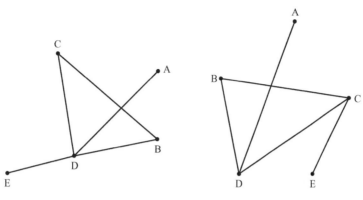

Figure 2.18　　　　　**Figure 2.19**

One way of establishing whether two graphs are isomorphic is to compare their adjacency matrices (allowing for the possibility of relabelling the vertices).

Example 2.4

Are the graphs with adjacency matrices shown in Tables 2.7 and 2.8 isomorphic?

	A	B	C
A	2	1	0
B	1	0	2
C	0	2	0

Table 2.7

	A	B	C
A	0	0	2
B	0	2	1
C	2	1	0

Table 2.8

Note

Any two complete graphs with *n* vertices are isomorphic.

Solution

By comparing the number of edges joining each pair of vertices, it can be seen that the graphs are isomorphic. The vertices have simply been relabelled.

Exercise 2.3

① Determine whether the graphs represented by the following adjacency matrices are isomorphic.

	A	B	C	D
A	0	0	1	2
B	0	0	1	1
C	1	1	2	0
D	2	1	0	2

Table 2.9

	A	B	C	D
A	2	1	0	1
B	1	0	2	0
C	0	2	2	1
D	1	0	1	0

Table 2.10

② Which of the following graphs are isomorphic?

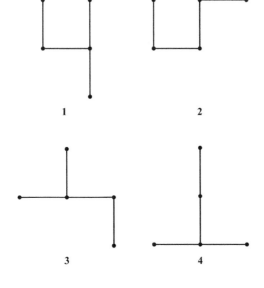

1 **2**

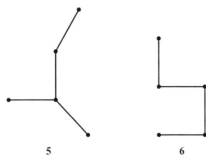

3 **4**

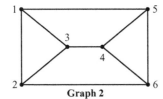

5 **6**

Figure 2.20

③ (i) Show that the following graphs are isomorphic.

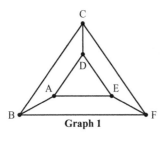

Graph 1

Graph 2

Figure 2.21

(ii) Draw a simply connected graph on six vertices, each of degree 3, which is not isomorphic to Graph 1/Graph 2.

④ Donald claims that the following graphs are isomorphic.

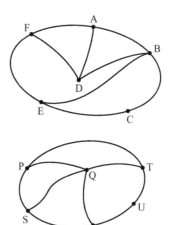

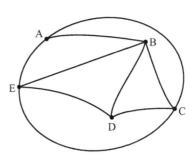

Figure 2.22

Explain why Donald is mistaken.

⑤ Prove that the following graphs are isomorphic.

Figure 2.23

	P	**Q**	**R**	**S**	**T**
P	0	1	1	1	1
Q	1	0	1	1	1
R	1	1	0	1	0
S	1	1	1	0	1
T	1	1	0	1	0

Table 2.11

4 Hamiltonian graphs

Another interesting problem is finding a route around a graph that visits all of the vertices exactly once. Note that edges cannot be repeated, as this would mean the repetition of a vertex (however, not all edges need to be traversed). You also need to be able to return to the starting point. If such a route exists, then it is called a **Hamiltonian cycle** (also known as a tour), and a graph that possesses a Hamiltonian cycle is called a Hamiltonian graph. (A **Hamiltonian path** visits all of the vertices exactly once, but may not be a cycle.) This idea is employed in the 'Travelling salesperson' problem that you will meet in Chapter 4.

ACTIVITY 2.3
Look back at Figure 2.1 again. Which town or city should be removed, in order to make the graph Hamiltonian?

Example 2.5

How many different Hamiltonian cycles does K_n have?

Solution

There will be $\frac{1}{2}(n-1)!$ possible Hamiltonian cycles: you can choose to start at any vertex, and there will be $n-1$ ways of choosing the next vertex to proceed to (and so on). You divide by 2 because reversing the order gives the same cycle.

A sufficient (though not necessary) condition for a graph to be Hamiltonian is given by **Ore's theorem**, which states that, for a simply connected graph G with $n \geq 3$ vertices, G will be Hamiltonian if $\deg v + \deg w \geq n$ for every pair of non-adjacent vertices.

Example 2.6

Use Ore's theorem to determine whether the graph in Figure 2.24 is Hamiltonian.

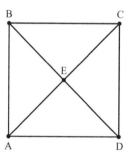

Figure 2.24

Solution

Referring to Figure 2.24, $\deg(A) + \deg(C) = 3 + 3 \geq 5$,

$\deg(B) + \deg(D) = 3 + 3 \geq 5$

So Ore's theorem $\Rightarrow$ the graph is Hamiltonian.

Example 2.7

Use Ore's theorem to determine whether the graph in Figure 2.25 is Hamiltonian.

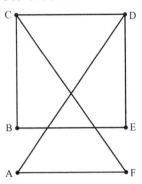

Figure 2.25

ACTIVITY 2.4

Show that the graph in Figure 2.25 is in fact Hamiltonian.

Solution

Referring to Figure 2.25, deg(A) + deg(B) = 2 + 2 = 4 < 6.

So we cannot use Ore's theorem to deduce that the graph is Hamiltonian.

Exercise 2.4

① Which of these graphs are Hamiltonian?

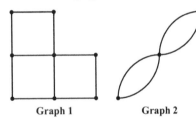

Graph 1   Graph 2

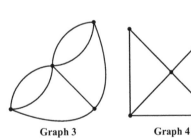

Graph 3   Graph 4

Figure 2.26

② Find a Hamiltonian cycle for this graph.

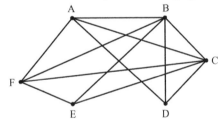

Figure 2.27

③ Regarding ABCDEA as different from AEDCBA, how many different Hamiltonian cycles are there in the graph in Figure 2.28?

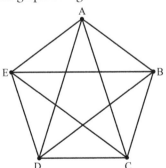

Figure 2.28

④ For each of the following graphs, establish what conclusion can be drawn from Ore's theorem, and whether the graph is Hamiltonian.

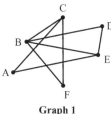

Graph 1

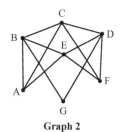

Graph 2

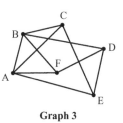

Graph 3

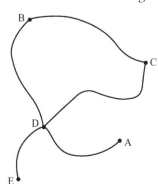

Graph 4

Figure 2.29

5 Planar graphs

A graph is said to be **planar** if it can be distorted in such a way that its edges do not cross. Figure 2.30 is planar, because it can be redrawn as Figure 2.31.

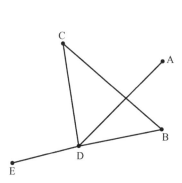

Figure 2.30

Figure 2.31

Alternative representations of a planar graph, such as those above, are isomorphic.

If a graph is both connected and planar, and if you consider a representation of the graph where the edges don't cross, then the plane containing the graph can be divided up into (regions or faces), which are bounded by the edges. You also include the infinite region with no boundary.

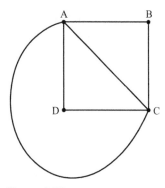

Figure 2.32

So the graph in Figure 2.32 has 4 regions, while the graph in Figure 2.31 has 2 regions.

Euler found that the following result holds for all connected planar graphs.

$$V + R = E + 2$$

where V, R and E are the numbers of vertices, regions and edges, respectively. This result is known as **Euler's formula**.

Example 2.8

Use Euler's formula to show that if the number of vertices in a connected planar graph is one greater than the number of edges, then the graph is a tree.

> **Solution**
>
> From Euler's formula, $V + R = E + 2$.
>
> Then, if $V = E + 1$, $R = 1$.
>
> If there is only one region, then there are no cycles, multiple edges or loops, and so the graph is a simply connected graph with no cycles; i.e. a tree.

The **thickness** of a graph is the minimum number of planar graphs that its edges can be grouped into. (This has applications to circuit boards in Electronics.)

Example 2.9

(i)   Find a Hamiltonian cycle for the graph below.

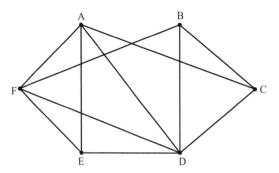

Figure 2.33

(ii) Starting with your cycle, add in edges to show that the original graph is planar.

Solution

(i)   A Hamiltonian cycle is shown in Figure 2.34.

(ii) For example, you can start with the cycle shown in Figure 2.34, and manoeuvre vertices B and C, in order to remove crossing edges (giving Figure 2.35). You then add in the necessary (non-crossing) edges to give Figure 2.36, which is a distorted version of the original graph, showing that the original graph is planar.

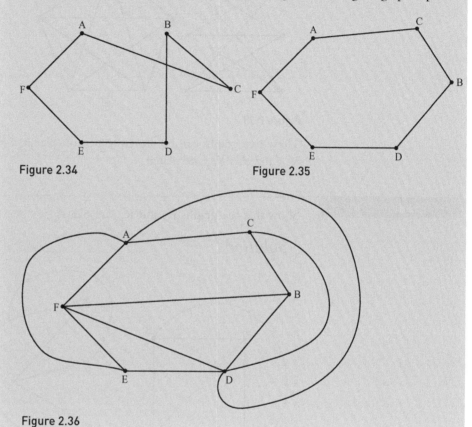

Figure 2.34

Figure 2.35

Figure 2.36

> **Note**
> ----------------
> Note that whilst a subgraph involves stripping out edges and (isolated) vertices, a subdivision involves adding in vertices and dividing edges into two.

A **subdivision** of a graph is obtained by inserting a new vertex into an edge (one or more times; zero times also counts as a subdivision). Figure 2.37 shows the effect of making two subdivisions of K_4 in Figure 2.24.

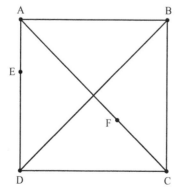

Figure 2.37

> **Discussion point**
> ➔ Is K_4 a subgraph of a subdivision of K_4?

An **edge contraction** occurs when an edge is removed and the two vertices that were joined to it are merged. Any edges that were incident to either of the original vertices now become incident to the new vertex.

Kuratowski's theorem

Kuratowski's theorem provides a way of determining whether a graph is planar.

The theorem involves the graphs $K_{3,3}$ and K_5 that are shown in Figure 2.38.

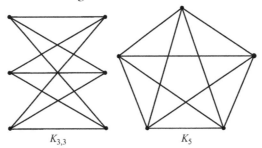

> Remember that a graph is said to be planar if it can be distorted in such a way that its edges do not cross.

$K_{3,3}$ K_5

Figure 2.38

These two graphs can be shown to be non-planar, whilst graphs such as K_4 and $K_{3,2}$, and also $K_{4,2}$, are planar.

Example 2.10

Show that the graphs K_4 and $K_{3,2}$ are planar.

Solution

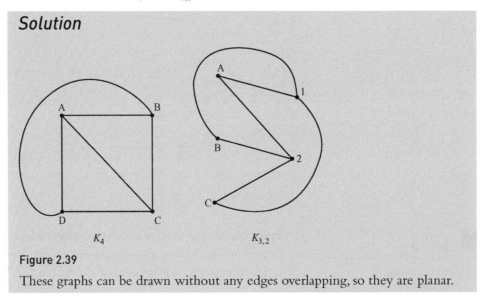

K_4 $K_{3,2}$

Figure 2.39

These graphs can be drawn without any edges overlapping, so they are planar.

A formal statement of Kuratowski's theorem will be given shortly, but for the moment you can say informally that any graph that is simpler than $K_{3,3}$ or K_5 will be planar.

This includes incomplete graphs, as in the following example.

Example 2.11

Show that if an edge is removed from K_5, then the resulting graph is planar.

Solution

Without loss of generality (since the graph is symmetric), you can remove the edge CE.

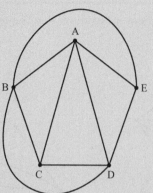

Figure 2.40

There are now no overlapping edges, so the graph is planar.

ACTIVITY 2.6

Show that if an edge is removed from $K_{3,3}$, then the resulting graph is planar.

The formal wording of Kuratowski's theorem is as follows.

A graph is non-planar if and only if it contains a subgraph that is a subdivision of either $K_{3,3}$ or K_5.

So to determine whether a graph is planar, you need to see if it has subgraphs that are subdivisions of either $K_{3,3}$ or K_5.

The examples that follow explain how to do this.

If the graph is bipartite, then you should generally be looking for $K_{3,3}$ as a subgraph; otherwise you will be looking for K_5.

> **Note**
>
> Recall that whilst a subgraph involves stripping out edges and (isolated) vertices, a subdivision involves adding in vertices and dividing edges into two. Also, note that $K_{3,3}$ and K_5 count as subdivisions of themselves.

Proof of Kuratowski's theorem

The theorem can be paraphrased informally as

a graph is non-planar if and only if it is as complex as $K_{3,3}$ or K_5.

There are two parts to the paraphrased theorem.

(i) A graph is non-planar if it is as complex as $K_{3,3}$ or K_5 (or 'as complex as $K_{3,3}$ or $K_5 \Rightarrow$ non-planar'; alternatively, 'planar $\Rightarrow$ not as complex as $K_{3,3}$ or K_5').

(ii) A graph is non-planar only if it is as complex as $K_{3,3}$ or K_5 (or 'non-planar $\Rightarrow$ as complex as $K_{3,3}$ or K_5'; alternatively, 'not as complex as $K_{3,3}$ or $K_5 \Rightarrow$ planar').

Part (i) sounds quite plausible, and is relatively easy to demonstrate. This will be done next. Part (ii) is less obvious, and the proof is beyond the scope of the A Level course.

To prove part (i), you need to show that if a graph G contains a subgraph that is a subdivision of either $K_{3,3}$ or K_5, then G is non-planar.

Suppose that G is planar. Then any subgraph of G will certainly be planar. But then suppose you are told that one such subgraph is a subdivision of either $K_{3,3}$ or K_5. If this subdivision is planar, then either $K_{3,3}$ or K_5 (as appropriate) would have to be planar, which gives a contradiction, as $K_{3,3}$ and K_5 are known to be non-planar. Therefore a graph that is as complex as $K_{3,3}$ or K_5 is non-planar.

Example 2.12

Use Kuratowski's theorem to show that the following graphs are non-planar.

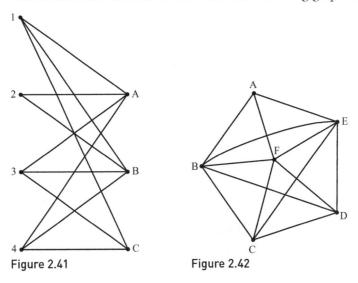

Figure 2.41   Figure 2.42

Solution

The bipartite graph in Figure 2.41 has $K_{3,3}$ (with vertices 1, 3, 4, A, B and C) as a subgraph. This is highlighted in Figure 2.43 with darker lines. Therefore, by Kuratowski's theorem, it is non-planar.

Similarly, the graph in Figure 2.42 has K_5 (with vertices B, C, D, E and F) as a subgraph. This is highlighted in Figure 2.44 with darker lines. Therefore it is also non-planar.

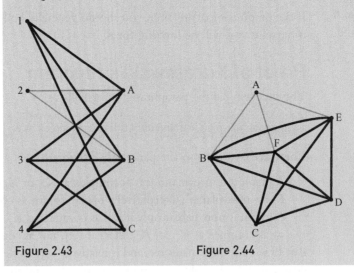

Figure 2.43   Figure 2.44

Example 2.13

Use Kuratowski's theorem to show that the following graphs are planar.

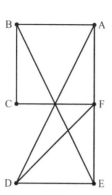

Figure 2.45

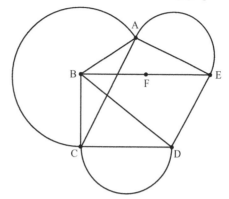

Figure 2.46

Solution

The graph in Figure 2.45 has bipartite graphs as subgraphs, but not $K_{3,3}$. Also, no combination of 5 vertices will produce K_5, so that K_5 is not a subgraph.

You also need to consider the possibility of there being a subdivision of either $K_{3,3}$ or K_5 within the graph.

Similarly, the graph in Figure 2.46 does not contain either $K_{3,3}$ or K_5 (even if you remove F).

Therefore, by Kuratowski's theorem, both graphs are planar.

Exercise 2.5

① Use Kuratowski's theorem to decide whether the following graphs are planar.

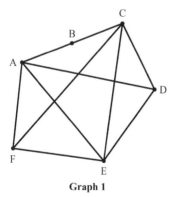

Graph 1

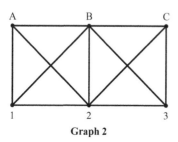

Graph 2

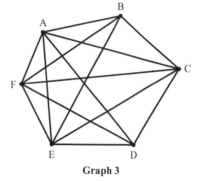

Graph 3

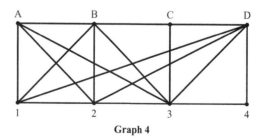

Graph 4

Figure 2.47

② Show that Euler's formula is satisfied for the two planar graphs in question 1.

③ Use Kuratowski's theorem to show that any graph with 8 or fewer edges is planar.

④ (i) Use Kuratowski's theorem to show that $K_{4,2}$ is planar, and confirm this by drawing the graph.

   (ii) Extend this to $K_{n,2}$.

⑤ Show that the graph in question 2, Exercise 2.4, is planar, by working from the Hamiltonian cycle discovered in that question.

⑥ Referring to Figure 2.48

   (i) Is it a subdivision of K_5?

   (ii) Is it a subgraph of K_5?

   (iii) Is it a subgraph of a subdivision of K_5?

   (iv) Does it contain a subgraph that is a subdivision of K_5?

   (v) Is it planar?

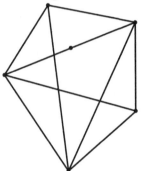

Figure 2.48

⑦ Determine whether the graph below is planar, fully justifying your answer.

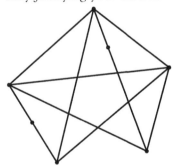

Figure 2.49

⑧ Vertices in a network are said to be adjacent if there is an edge joining them. Colours $c_1, c_2, c_3,$ ... are to be assigned to the vertices ($a, b, c,$ ...) of a graph so that no two adjacent vertices share the same colour. A method for doing this is as follows:

   1 Allocate colour c_1 to vertex a.

   2 Choose the next uncoloured vertex in alphabetical order. List the colours of all the vertices adjacent to it that are already coloured. Choose the first colour that is not in that list and allocate that colour to the vertex.

3 Repeat Step 2 until all vertices are coloured.

(i) (a) Use the method to complete the colouring of the following network, taking the vertices in alphabetical order.

   Indicate the colours that you allocate ($c_1, c_2,$ etc.) in the boxes provided.

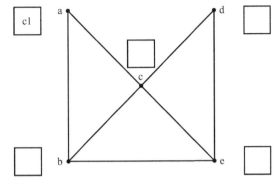

Figure 2.50

   (b) Give a colouring that uses fewer colours.

(ii) (a) Explain why it is not possible to colour vertices in networks with loops.

   (b) Explain why repeated edges can be ignored when colouring vertices in a network.

(iii) At a mathematics open day, 6 one-hour lectures a, b, c, d, e and f are to be scheduled so as to allow students to choose from the following combinations.

a & b   a & d   c & e
b & f   d & e   e & f   a & f

   (a) Represent this as a network, connecting two vertices with an edge if they must not be scheduled at the same time.

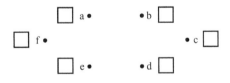

Figure 2.51

   (b) Use the method to colour the network, taking the vertices in alphabetical order. Show the colours that you allocate ($c_1, c_2,$ etc.) on your network.

   (c) Say how the colouring can be used to give a schedule for the lectures, and give the schedule implied by your colouring.

   (d) Find a best schedule for the lectures.

[MEI adapted]

There are six people in a room. Prove that there must be at least three people who know each other, or at least three people none of whom know one another.

 There are a number of people at a party, and some handshaking has taken place. Prove that there must be two people who have shaken hands with the same number of people.

LEARNING OUTCOMES

Now you have finished this chapter, you should be able to
➤ understand and use the language of graphs, including vertex, node, edge, arc, network, trail, cycle, connected, degree, subgraph, subdivision, multiple edge and loop
➤ identify or prove properties of a graph, including that a graph is Eulerian, semi-Eulerian or Hamiltonian
➤ understand and use Ore's theorem
➤ understand and use Euler's formula for connected planar graphs
➤ understand and use complete graphs and bipartite graphs, including adjacency matrices
➤ understand and use simple graphs, simply connected graphs and trees
➤ use Kuratowski's theorem to determine the planarity of graphs
➤ recognise and find isomorphism between graphs.

KEY POINTS

1 A graph consists of vertices and edges.
2 A trail is a sequence of edges in which the end of one edge is the start of the next, and where no edge is repeated.
3 A path is a trail with the further restriction that no vertex is repeated.
4 A cycle is a closed path.
5 A graph is connected if there exists a path between every pair of vertices.
6 A simple graph is one that has no multiple edges or loops.
7 A tree is a simply connected graph with no cycles.
8 The degree of a vertex is the number of edges that join it.
9 A graph is Eulerian if it contains a closed trail that includes all the edges. It is semi-Eulerian if there is a trail that includes all the edges, but is not closed.
10 A graph is Hamiltonian if it contains a cycle that visits all of the vertices exactly once.
11 Ore's theorem says that, for a simply connected graph G with $n \geq 3$ vertices, G will be Hamiltonian if $\deg v + \deg w \geq n$ for every pair of non-adjacent vertices.
12 A graph is planar if it can be distorted in such a way that its edges do not cross.
13 Euler's formula: $V + R = E + 2$
14 An adjacency matrix shows the number of edges connecting any two vertices.
15 A complete graph is one where every two vertices share exactly one edge (and where there are no loops).
16 A subdivision of a graph is obtained by inserting a new vertex into an edge.
17 A bipartite graph contains two sets of vertices, with edges only joining a vertex in one set to a vertex in the other.
18 Two graphs are isomorphic if one can be distorted in some way to produce the other.
19 Kuratowski's theorem says that a graph is non-planar if and only if it contains a subgraph that is a subdivision of either $K_{3,3}$ or K_5.

3

Algorithms

An algorithm must be seen to be believed.
Donald Knuth (1938–)

→ Use long division to work out 98 606 ÷ 47.

1 What is an algorithm?

An algorithm is a finite sequence of operations for carrying out a procedure or solving a problem. Cooking recipes, knitting patterns and instructions for making flat pack furniture are algorithms, but obviously these are not mathematical algorithms.

ACTIVITY 3.1

Do you know on which day of the week you were born? Zeller's algorithm can be used to work it out. Try the algorithm using your date of birth.

Zeller's algorithm	Example: 29th Feb 2000
Let day number = D Let month number = M Let year number = Y	D = 29 M = 2 Y = 2000
If M = 1 or 2, add 12 to M and subtract 1 from Y	M = 14 Y = 1999
Let C be the first two digits of Y and X be the last two digits of Y	C = 19 X = 99
Calculate $\text{INT}(2.6M - 5.4) + \text{INT}(X \div 4) + \text{INT}(C \div 4) + D + X - 2C$	31 + 24 + 4 + 29 + 99 − 38 = 149
Find the remainder when this is divided by 7	2
If the remainder is 0 the day was Sunday, if it is 1 the day was Monday, and so on	Tuesday

Table 3.1

This uses the current (new) value of Y.

$\text{INT}(N)$ is the integer part of N. This is the largest integer that is less than or equal to N. For example, $\text{INT}(2.3) = 2$, $\text{INT}(6.7) = 6$, $\text{INT}(4) = 4$ and $\text{INT}(0) = 0$. The integer part of a negative number N is the negative of $\text{INT}(-N)$, for example, $\text{INT}(-1.7) = -1$.

The remainder when N is divided by 7 is the same as $N - 7 \times \text{INT}(N \div 7)$

In mathematics an algorithm has an initial state and involves **inputs**, **outputs** and variables. The initial state is the 'factory setting' values of any variables that are not defined within the algorithm. Usually this means that all variables have the value 0 until they are updated.

For Zeller's algorithm as given above, the inputs are the initial values of D, M and Y; the output is the day and the variables are C, D, M, X and Y.

The output may be printed or displayed.

Communicating an algorithm

How do you communicate an algorithm? The form of communication depends on who (for example a seven-year-old) or what (for example a computer) will be using the algorithm.

Whatever method is used to communicate the steps of an algorithm it must be:
- **deterministic**, so there is no chance involved and the same input will always produce the same output
- **finite**, so the algorithm stops.

An algorithm may be communicated in several ways:
- ordinary language, as in Zellor's algorithm in Activity 3.1
- a flowchart, as in the solution to Example 3.1 on page 38
- pseudo-code, as follows.

Zeller's algorithm could be written in pseudo-code as:

Step 1   Let D = day number
         Let M = month number
         Let Y = year number

M = M + 12 means that the value of (old)
M + 12 becomes the value of (new) M.

Step 2   If M < 3 then M = M + 12 and Y = Y − 1

Step 3   Let C = INT(Y ÷ 100)
         Let X = Y − (100 × C)

Step 4   Let S = INT(2.6M − 5.4) + INT(X ÷ 4) + INT(C ÷ 4) + D + X − 2C

Step 5   Let A = S − (7 × INT(S ÷ 7)) and display the value of A

The output is the day number and not the name of the day. However, it is easy to adapt the algorithm so that it converts the number to the day and outputs that. The algorithm is worked through once, with each step being performed once. Some algorithms involve decisions ('if … then…') and may loop back to an earlier step or another point in the algorithm ('go to Step…'). In this case, the algorithm will continue indefinitely unless there is a stopping mechanism incorporated into the algorithm.

For example, an algorithm for finding square roots is given below:

Step 1   Input a positive number N

Step 2   Let A = $\frac{1}{2}$N

ACTIVITY 3.2

Work through the algorithm with N = 2.

Step 3   Let $B = \frac{1}{2}\left(A + \frac{N}{A}\right)$

This is the condition for stopping the algorithm. → Step 4   If $(A - B)^2 < 0.001$ then go to Step 6

Step 5   Let $A = B$ and then go to Step 3

This stops the algorithm. → Step 6   Display the value of B and STOP

Example 3.1

The real roots of a quadratic equation

$$ax^2 + bx + c = 0 \ (a \neq 0)$$

can be found using the quadratic formula

$$x = \frac{-b \pm \sqrt{b^2 - 4ac}}{2a}$$

(i)   Use a flowchart to represent an algorithm for solving a quadratic equation.

(ii)   Write the algorithm in pseudo-code.

Solution

(i)

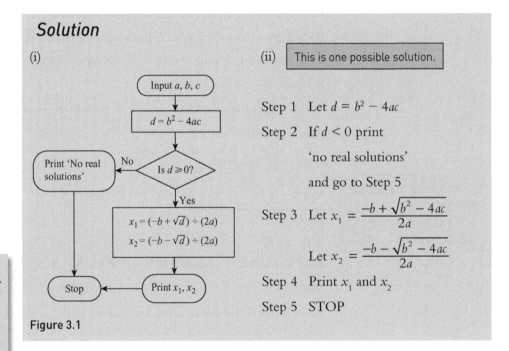

(ii)   This is one possible solution.

Step 1   Let $d = b^2 - 4ac$

Step 2   If $d < 0$ print

'no real solutions'

and go to Step 5

Step 3   Let $x_1 = \dfrac{-b + \sqrt{b^2 - 4ac}}{2a}$

Let $x_2 = \dfrac{-b - \sqrt{b^2 - 4ac}}{2a}$

Step 4   Print x_1 and x_2

Step 5   STOP

Figure 3.1

> ### Note
> The word 'algorithm' has become more commonplace since the development of the computer. A computer program is simply an algorithm written in such a way that a machine can carry it out.

An algorithm gives the logical structure that underlies a computer program for solving a problem. Using algorithms has obvious connections with computer science but does not require programming skills or knowledge of any specific computing language.

ACTIVITY 3.3

Below are two algorithms, one expressed in pseudo-code and the other as a flowchart.

Russian algorithm for multiplying two integers

Step 1   Write the two numbers side by side

Step 2   Beneath the left number write double that number

Beneath the right number write the integer part of half that number

Step 3   Repeat Step 2 until the right number is 1

Step 4   Delete those rows where the number in the right column is even

Step 5   Add up the remaining numbers in the left column.

This is the result of multiplying the original numbers

Euclid's method for finding the highest common factor of two positive integers x and y

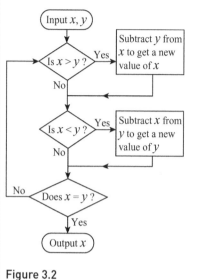

Figure 3.2

Work through the Russian algorithm with left number 13 and right number 37.

Work through the Euclidean algorithm with $x = 6$ and $y = 15$.

How could you represent these algorithms differently?

What different type of user might each representation be suitable for?

When you worked through the algorithms in Activity 3.3 you probably kept track of the values at each stage. When this is done rigorously, it is called **tracing** the algorithm.

Example 3.2

Here is an algorithm expressed in pseudo-code.

Step 1   N = 0; Input K

Step 2   For J = 1 to K

Step 3   Input x, the next value in the list

Step 4   If ABS(x) = x then go to Step 6

Step 5   N = N + 1

Step 6   Next J

Step 7   Print N

Step 8   Stop

(i)   Trace the algorithm for K = 5 and x inputs $-2, 4, -6, 9, -1$.

(ii)   What does the algorithm do?

(iii) What functions do Steps 1, 5 and 7 perform in the algorithm?

(iv)   Could the algorithm be improved?

> ## Solution
>
> (i)   Trace the values of N, J and x.
>
> | N | J | x |
> |---|---|-----|
> | 0 | | |
> | 1 | 1 | -2 |
> | | 2 | 4 |
> | 2 | 3 | -6 |
> | | 4 | 9 |
> | 3 | 5 | -1 |
>
> 3 is printed at the end
>
> Table 3.2
>
> (ii)   It counts the number of negative numbers in the list.
>
> (iii) Step 1 starts the counter at zero.
>
> Step 5 increases the counter by one each time a negative number is encountered.
>
> Step 7 prints the total number of negative numbers in the list.
>
> (iv)   Change Step 4 to
>
> If $ABS(x) \neq x$ then N = N + 1
>
> and remove Step 5 so there are fewer steps altogether.

> ### Note
>
> $ABS(x)$ returns the magnitude of x. $ABS(-5) = 5$. It means 'absolute value of'.

There is one variable, J, counting the items in the list of inputs, and another, N, counting the negative numbers. Both of these are examples of **counters**.

Counters are used in many algorithms, as you will see in the rest of this chapter.

An algorithm for calculating $n!$ using a counter might look like this:

Step 1   K = 1, Input N ($N \in \mathbb{Z}, N \geqslant 2$)

Step 2   For J = 2 to N

Step 3   K = K*J

Step 4   Next J

Step 5   Output K

This is an iterative algorithm. An alternative way of calculating $n!$ is to use a **recursive** algorithm. Instead of building up the value from 1, the algorithm uses $n! = n \times (n - 1)!$ to repeatedly call up a simpler version until it reaches a base case that is known, $0! = 1$.

Algorithms do not always need to be written from scratch. In many cases an existing algorithm can be adapted to do the job, or a set of instructions is available to work from.

Example 3.3

Numbers in base 10 can be converted to binary (base 2). To do this you keep dividing by two, recording the remainder each time until you reach a quotient of 1. Then you write down the final quotient and each of the remainders in reverse order, including the zeros.

(i) Rewrite the instructions as an algorithm to convert a number in base 10 to binary.

(ii) Adapt your algorithm to convert a number in base 10 to ternary (base 3).

Solution

(i) Step 1   Input N

Step 2   Divide N by 2 and record remainder to the left of any existing remainders

Step 3   Quotient becomes N

Step 4   If N > 1 go to Step 2

Step 5   Record the value of N to the left of list of remainders

Step 6   Print list of remainders

(ii) Change Step 2 to 'Divide N by 3 and record remainder to the left of any existing remainders' and change Step 4 to say 'If N > 2 go to Step 2'

Exercise 3.1

① Construct a flowchart that can be used to check if a number N is prime, where N is a positive integer and N > 2.

② The following six steps define an algorithm:

Step 1   Think of a positive whole number and call it X

Step 2   Write X in words (using letters)

Step 3   Let Y be the number of letters used

Step 4   If $Y = X$ then stop

Step 5   Replace X by Y

Step 6   Go to Step 2

(i) Apply the algorithm with $X = 62$.

(ii) Show that for all values of X between 1 and 99 the algorithm produces the same answer.

You may use the fact that, when written out, numbers between 1 and 99 all have twelve or fewer letters.   [MEI]

③ The following instructions operate on positive integers greater than 4.

Step 10   Choose any positive integer greater than 4, and call it n

Step 15   Write down n

Step 20   If n is even then let $n = \dfrac{n}{2}$ and write down the result

Step 30   If n is odd then let $n = 3n + 1$ and write down the result

Step 40   Go to Step 20

(i) Apply the instruction with 6 as the chosen integer, stopping when a sequence repeats itself.

(ii) Apply the instructions with 256 as the chosen integer, stopping when a sequence repeats itself.

(iii) Add an instruction to stop the process when n becomes 1.

(iv) It is not known if, when modified to stop cycling through 4, 2, 1, the instructions form an algorithm. What would need to be known for it to be an algorithm? [MEI]

④ A bag contains 26 cards. A different letter of the alphabet is written on each one. A card is chosen at random and its letter is written down. The card is returned to the bag. The bag is shaken and the process is repeated several times.

Tania wants to investigate the probability of a letter appearing twice. She wants to know how many cards need to be chosen for this probability to exceed 0.5. Tania uses the following algorithm.

Step 1 Set $n = 1$

Step 2 Set $p = 1$

Step 3 Set $n = n + 1$

Step 4 Set $p = p \times \dfrac{27 - n}{26}$

Step 5 If $p < 0.5$ then stop

Step 6 Go to Step 3

(i) Run the algorithm.

(ii) Interpret your results.

A well-known problem asks how many randomly-chosen people need to be assembled in a room before the probability of at least two of them sharing a birthday exceeds 0.5 (ignoring anyone born on 29 February).

(iii) Modify Tania's algorithm to answer the birthday problem. (Do not attempt to run your modified algorithm.)

(iv) Why have 29 February birthdays been excluded? [MEI]

⑤ Table 3.3 can be used to convert a number from Roman numerals into ordinary base 10 numbers.

Row	M		D		C		L		X		V		I	
1	1000	2	500	3	100	9	50	5	10	10	5	7	1	11
2	1000	2	500	3	100	9	50	5	10	10	5	7	1	11
3					100	9	50	5	10	10	5	7	1	11
4					100	4	50	5	10	10	5	7	1	11
5							50	6	10	10	5	7	1	11
6									10	6	5	7	1	11
7											5	8	1	11
8													1	8
9	800	5	300	5	100	4	50	6	10	10	5	8	1	11
10					80	7	30	7	10	6	5	8	1	11
11									8	0	3	0	1	8

Table 3.3

To illustrate how this works, take the Roman numeral CIX as an example.

Always start by looking at row 1. Look at the row 1 entry in the column headed C (the first symbol in the Roman numeral) to find 100 9. Add 100 to the running total (which was 0 originally) and move to row 9.

Now look at row 9 in the column headed I (the second symbol in the Roman numeral) to find 1 11. Add 1 to the running total and move to row 11.

Finally look at row 11 in the column headed X (the third symbol in the Roman numeral) to find 8 0. Add 8 to the running total. Since this was the last symbol in the Roman numeral the algorithm now stops.

CIX = 100 + 1 + 8 = 109

(i) Write this algorithm as a set of steps.

(ii) What are the limitations of the algorithm?

(iii) Write pseudo-code instructions for converting ordinary base 10 numbers into Roman numerals.

[MEI adapted]

2 Algorithmic complexity

Most problems can be solved using a variety of algorithms, some of which might be more efficient than others. By 'efficient' we usually mean using fewer operations (which in turn means running more quickly and so taking less time). There might be other considerations too, such as the amount of storage capacity needed if the algorithm is to be run on a computer.

As a simple example of improving efficiency, look back to Example 3.1, where you saw an algorithm to find the real roots of a quadratic equation. It is a good idea to calculate the value of $b^2 - 4ac$ as a first step, because the sign of that value has to be checked to see whether it is worth continuing with the calculation.

If an algorithm requires the evaluation of a quadratic expression, the way in which the expression is written can make a difference to the efficiency.

For example, $3x^2 + 2x + 9$ can be written as $(3x + 2)x + 9$. This bracketed form is called a nested form.

When $x = 5$, the evaluation with a calculator requires the following key presses:

$3x^2 + 2x + 9$: $(\ 3\ \times\ 5\ \times\ 5\ )\ +\ (\ 2\ \times\ 5\ )\ +\ 9\ =$ 3 multiplications and 2 additions $(3x + 2)x + 9$: $(\ 3\ \times\ 5\ +\ 2\ )\ \times\ 5\ +\ 9\ =$ 2 multiplications and 2 additions

The nested form uses fewer operations (multiplications and additions) so it should be quicker (albeit by the tiniest amount of time).

Comparing the number of operations for a general polynomial of degree n gives:

$a_nx^n + a_{n-1}x^{n-1} + \ldots + a_2x^2 + a_1x + a_0$	$\frac{1}{2}n(n + 1) \times$ and $n +$
$(((\ldots((a_nx + a_{n-1})x + a_{n-2})x + \ldots) + a_1)x + a_0$	$n \times$ and $n +$

> **Note**
> ---
> $\frac{1}{2}n(n + 1)$ is the sum of the first n natural numbers. It occurs in many calculations of complexity.

The nested method has **linear order complexity** (or order n or $O(n)$) because the time taken to run the calculation involves n^1 as the highest power of n.
The expanded form has **quadratic order complexity** (or order n^2 or $O(n^2)$) because the time taken to run the calculation will involve n^2 as the highest power of n.

The nested method is more efficient than the expanded form because $O(n)$ is a lower order complexity than $O(n^2)$. Irrespective of what the actual linear and quadratic functions are that represent the run-time for the two methods for a polynomial of degree n, a linear function will give lower values (smaller run-times) than a quadratic function for realistic sizes of (huge) problems.

> **Note**
> ---
> If any of the coefficients happened to be 0 then some work would be saved. It is usual to focus on the worst case situation (rather than the best case or an average case). This is partly because then any predictions about run-times will be 'worst case scenarios', but mainly because the worst case is usually the easiest to consider.

For example, if it takes M microseconds for a computer to multiply two numbers and A microseconds for it to add two numbers then (once the programs have been written) inputting the coefficients is the same for both methods and the run-time is $\frac{1}{2}n(n + 1)M + nA$ for the expanded form and $nM + nA$ for the nested form.

Now, in this case, $\frac{1}{2}n(n + 1)M + nA$ is always bigger than $nM + nA$, but the details, such as the $\frac{1}{2}$ and the $(n + 1)$ are irrelevant: if n is huge, all that matters is that n^2 is much bigger than n.

$O(n^2)$ is always less efficient than $O(n)$ once n becomes large. Similarly $O(n^3)$ is less efficient than $O(n^2)$ and so on.

- If an algorithm has $O(n)$ complexity then doubling the size of the problem will roughly double the run-time, or tripling the size of the problem will roughly triple the run-time. If the actual run-time is $an + b$ then scaling the problem size by a factor of k gives a run-time of $akn + b$. For large values of n, the runtime $akn\ (+ b)$ is roughly k times the original run-time $an\ (+ b)$.

- If an algorithm has $O(n^2)$ complexity then doubling the size of the problem will roughly quadruple the run-time, or tripling the size of the problem will scale the run-time by a factor of about 9. If the actual run-time is $an^2 + bn + c$ then scaling the problem size by a factor of k gives a run-time of $ak^2n^2 + bkn + c$.

For large values of n, the run-time $ak^2n^2 (+ bkn + c)$ is roughly k^2 times the run-time $an^2 (+ bn + c)$.

- Similarly, if an algorithm has $O(n^r)$ complexity, then scaling the problem size by a factor k will scale the run-time by a factor of (approximately) k^r.

Example 3.4

An algorithm has $O(n^2)$ complexity. A problem, using this algorithm, has run time, $T(n)$, of 0.02 seconds when $n = 40$. What is the approximate run-time when $n = 200$?

Note

Make sure you know the difference between run-time, $T(n)$, and the order of an algorithm.

Solution

The algorithm has complexity of $O(n^2)$; so, for large values of n, the scale factor for the run-time is the square of the scale factor for the size of the problem.

The scale factor for the size $= \dfrac{200}{40} = 5$; the scale factor for the run-time $= 5^2 = 25$.

So the new run-time $= 25 \times 0.02 = 0.5$ seconds.

Exercise 3.2

① (i) An algorithm has linear complexity. It takes 4 milliseconds for it to solve a problem of size $n = 20$. Approximately how long does it take to solve a similar problem with size $n = 600$?

(ii) An algorithm has complexity $0(n^4)$. A problem with size 30 takes it 0.004 seconds to solve. Estimate the run-time for a similar problem with size 900.

(iii) The complexity of an algorithm is $0(n^3)$. A problem with $n = 6000$ takes it 2 seconds to solve. Why can't you say that the run-time for a similar problem of size $n = 6$ is 2×10^{-9} seconds?

② A particular algorithm has a run-time of 0.04 seconds when solving a problem of size $n = 25$. It has a run-time of 0.64 seconds when solving a similar problem of size $n = 100$. What can you deduce about the complexity of the algorithm?

③ The following flowchart defines an algorithm which operates on two inputs, x and y.

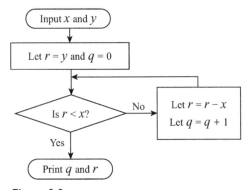

Figure 3.3

(i) Run the algorithm with inputs of $x = 3$ and $y = 41$. Keep a record of the values of r and q each time they are updated.

(ii) Say what the algorithm achieves.

The following flowchart defines an algorithm with three inputs, x, y_1 and y_2.

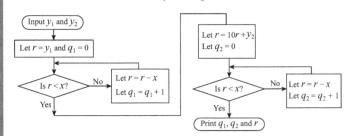

Figure 3.4

(iii) Work through the algorithm with $x = 3$, $y_1 = 4$ and $y_2 = 1$. Keep a record of the values of r, q_1 and q_2 each time they are updated.

The two algorithms achieve the same result.

(iv) Suggest the advantages and disadvantages of each algorithm. [MEI]

④ Programmable calculators use a version of the Basic programming language that, amongst other things, can perform repetitions using 'for … next'.

To show how this works look at the following programs and their printouts.

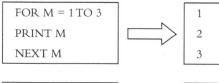

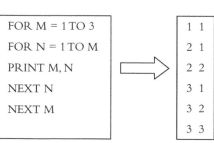

In a certain town the bus tickets are numbered 0000 to 9999.

Some children are collecting the tickets whose digits add up to 21.

(i) How many such tickets will there be in the tickets numbered from 0000 to 9999?

Two algorithms for finding the number of tickets whose digits add up to 21 are given as calculator programs below (A and B). You do NOT need to put these programs into your calculator to answer this question.

(ii) Show that each algorithm achieves the correct result.

(iii) Compare the efficiency of the two algorithms by counting the number of additions/subtractions and the number of comparisons used.

A

```
T = 0
FOR J = 0 TO 9
FOR K = 0 TO 9
FOR L = 0 TO 9
FOR M = 0 TO 9
S = J + K + L + M
IF S = 21 THEN T = T + 1
NEXT M
NEXT L
NEXT K
NEXT J
PRINT T
```

B

```
T = 0 : S = 0
FOR J = 0 TO 9
FOR K = 0 TO 9
FOR L = 0 TO 9
FOR M = 0 TO 9
IF S = 21 THEN T = T + 1
S = S + 1
NEXT M
S = S − 9
NEXT L
S = S − 9
NEXT K
S = S − 9
NEXT J
PRINT T
```

⑤ The following algorithm finds the highest common factor (HCF) of two positive integers.

1   Let A be the first integer and B be the second integer
2   Let Q = INT(B ÷ A)
3   Let R = B − (Q × A)
4   If R = 0 go to Step 8
5   Let the new value of B be A
6   Let the new value of A be R
7   Go to Step 2
8   Record the HCF as the value of A
9   STOP

(i) Work through the algorithm with A = 2520 and B = 5940.

(ii) What happens if the order of the input is reversed, so A = 5940 and B = 2520?

It has been claimed that the number of iterations of this algorithm is approximately

$$\dfrac{\log(M \div 1.17)}{\log\left(\dfrac{1 + \sqrt{5}}{2}\right)},$$ where M is the larger of A and B.

(iii) Count the number of iterations of the loop in the algorithm when A = 233 and B = 377.

Compare this with the number claimed by the formula above.

3 Packing

One of the situations where an algorithmic approach is useful is for one-dimensional bin-packing problems. Imagine having to send a number of files as attachments to the same email address, but there is a limit on the total size of the files attached to one email. How should the files be put together so that the number of emails needed is as small as possible?

The classic bin-packing problem packs 'boxes' of given sizes into a number of (equal sized) 'bins'.

Here are four methods that could be used:

1   **Next-fit algorithm**
    Take the boxes in the order listed and pack each box in the first bin that has enough space for it (starting with the current bin, where the most recent box was packed).

2   **First-fit algorithm**
    Take the boxes in the order listed and pack each box in the first bin that has enough space for it (starting with the first bin each time).

3   **First-fit decreasing algorithm**
    Reorder the boxes from the largest to the smallest, then apply the first-fit method to this list.

4   **Full-bin strategy**
    Look for combinations of boxes that fill bins. Pack these boxes. Put the rest together in combinations that result in bins that are as nearly full as possible.

Example 3.5

The boxes A to K with masses in kilograms as shown in Table 3.4 are to be packed into bins that can each hold a maximum of 15 kg. Apply each of the four bin packing methods to this problem.

A	B	C	D	E	F	G	H	I	J	K
8	7	4	9	6	9	5	5	6	7	3

Table 3.4

Solution

Next-fit:

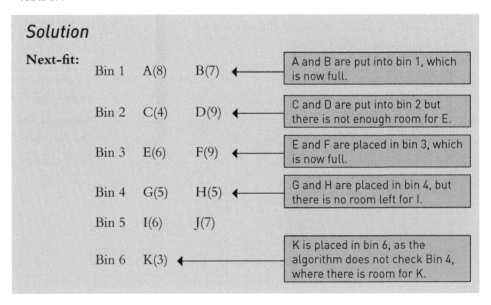

Bin 1   A(8)   B(7)   ← A and B are put into bin 1, which is now full.

Bin 2   C(4)   D(9)   ← C and D are put into bin 2 but there is not enough room for E.

Bin 3   E(6)   F(9)   ← E and F are placed in bin 3, which is now full.

Bin 4   G(5)   H(5)   ← G and H are placed in bin 4, but there is no room left for I.

Bin 5   I(6)   J(7)

Bin 6   K(3)   ← K is placed in bin 6, as the algorithm does not check Bin 4, where there is room for K.

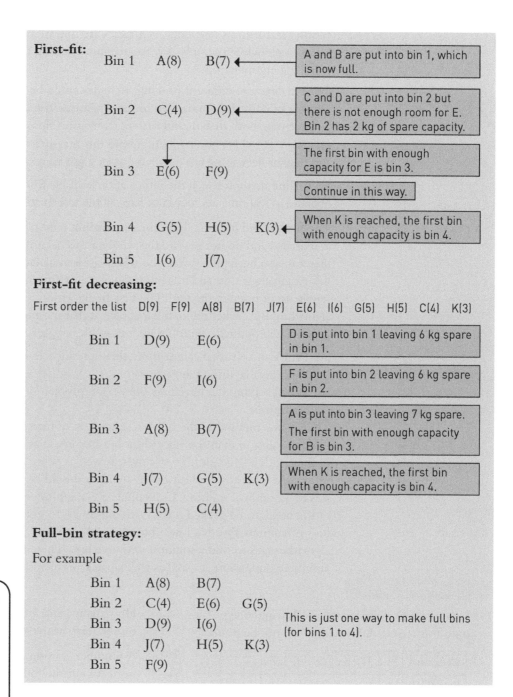

First-fit:

Bin 1	A(8)	B(7)

> A and B are put into bin 1, which is now full.

Bin 2	C(4)	D(9)

> C and D are put into bin 2 but there is not enough room for E. Bin 2 has 2 kg of spare capacity.

Bin 3	E(6)	F(9)

> The first bin with enough capacity for E is bin 3.

> Continue in this way.

Bin 4	G(5)	H(5)	K(3)

> When K is reached, the first bin with enough capacity is bin 4.

Bin 5	I(6)	J(7)

First-fit decreasing:

First order the list D(9) F(9) A(8) B(7) J(7) E(6) I(6) G(5) H(5) C(4) K(3)

Bin 1	D(9)	E(6)

> D is put into bin 1 leaving 6 kg spare in bin 1.

Bin 2	F(9)	I(6)

> F is put into bin 2 leaving 6 kg spare in bin 2.

Bin 3	A(8)	B(7)

> A is put into bin 3 leaving 7 kg spare. The first bin with enough capacity for B is bin 3.

Bin 4	J(7)	G(5)	K(3)

> When K is reached, the first bin with enough capacity is bin 4.

Bin 5	H(5)	C(4)

Full-bin strategy:

For example

Bin 1	A(8)	B(7)	
Bin 2	C(4)	E(6)	G(5)
Bin 3	D(9)	I(6)	
Bin 4	J(7)	H(5)	K(3)
Bin 5	F(9)		

> This is just one way to make full bins (for bins 1 to 4).

Discussion points

- → What does optimal mean in this sense?
- → Why might some other criterion be more appropriate?
- → What are the problems with using a complete enumeration?

The full-bin strategy is not an algorithm because there is no set way to make the full bins. It is probably evident that using the items with the largest weights to make full bins is usually better than filling a bin with lots of smaller weight items.

The only known algorithm that will always find the optimal packing is to use a complete enumeration (that is, try every possibility).

An algorithm that will usually find a good solution, although not necessarily an optimal (best) solution to a problem is called a **heuristic** (or a heuristic algorithm). A heuristic is a method that finds a solution efficiently, but with no guarantee that the solution is optimal. This raises another issue in the consideration of what constitutes an efficient algorithm. If you cannot guarantee getting the best solution would you want the algorithm that got to its solution with the least expense (in terms of time or money) or the one that consistently gave better solutions than the others? The consistency with which an algorithm gives a good solution is therefore another factor in its efficiency. Heuristics are important when classic methods fail, for example when the only way to guarantee finding the optimal solution is a complete enumeration.

The efficiency of different packing strategies could be compared by counting the number of comparisons needed in the worst case for a list of n items. In the worst possible case, both first-fit and first-fit decreasing algorithms use $1 + 2 + \ldots + (n-1) = \frac{1}{2}(n-1)n$ comparisons. Using this measure of complexity, both first-fit and first-fit decreasing are quadratic order algorithms, $O(n^2)$.

An **offline** method has all the information available from the start. An **online** method operates in real time and does not have all the information available at the start.

The first-fit and next-fit algorithms are **online** methods as they deal with each item as it arrives and do not require the whole list to be available at the start. Clearly they can also be applied when the whole list is available at the start. The first-fit decreasing algorithm and full-bin strategy are **offline** methods as they require the whole list of items at the start. The first-fit algorithm is an example of a **greedy** algorithm, as it makes the optimal choice at that moment but it does not necessarily generate the optimal solution. Other examples of greedy algorithms are

(i) Prim's and Kruskal's algorithms; these produce an optimal solution using the best available option at each step
(ii) When solving the travelling salesperson problem using the 'nearest neighbour' algorithm.

The ad-hoc method of the full-bin strategy is, in fact, more likely to result in an optimal solution than the algorithms but becomes very time-consuming when dealing with a large number of boxes, as well as being an offline method, and so unsuitable for many applications. A solution should be checked to see how close it is to an optimal solution. In Example 3.5 the bins contain 15 items and the sum of the boxes to be packed is 69. Dividing 69 by 15 gives 4.6 so a minimum of five bins is required. Five is a lower bound for the optimal solution. Only the next-fit algorithm fails to find a solution with five bins. These ideas are extended to two and three dimensions, as well as to knapsack problems, in Section 5.

Exercise 3.3

① Sam wants to download the following videos onto four 16 GB USB sticks. Can this be done?

Program	A	B	C	D	E	F	G	H	I
Size (GB)	4	3.2	2.4	2.6	4.4	1	2	2.4	3.6
Program	J	K	L	M	N	O	P	Q	R
Size (GB)	5.6	3	4.8	3	4	2.8	8	4.8	1.6

Table 3.5

② A small car ferry has a number of lanes, each 20 m long. The following vehicles are waiting to be loaded.

Petrol tanker	14 m	Car	4 m	Range rover	5 m	Car	4 m
Car	3 m	Van	4 m	Car and trailer	8 m	Car	3 m
Coach	12 m	Lorry	11 m	Car		4 m	Lorry	10 m

Table 3.6

How many lanes are needed to fit all the vehicles on the ferry at the same time?

③ A plumber is using pipes that are 6 m long and needs to cut the following lengths.

Length (m)	0.5	1	1.5	2	2.5	3	3.5
Number	0	2	4	3	0	1	2

Table 3.7

Use the first-fit decreasing algorithm to find a way to cut the lengths.

④ Six items with the masses given in Table 3.8 are packed into bags, each of which has a capacity of 10 kg.

Item	A	B	C	D	E	F
Weight (kg)	2	1	6	3	3	5

Table 3.8

(i) Use the first-fit algorithm to pack these items into bags, saying how many bags are needed.

(ii) Give an optimal solution. [MEI]

⑤ The coach of a netball team has to arrange three pre-season training sessions, each of length 90 minutes. She wants to schedule the activities that are listed below. Some are to be scheduled more than once.

Activity	Duration (mins)	Number of times activity is to be scheduled
A shooting practice	10	3
B passing practice	15	3
C blocking practice	12	3
D sprinting	5	3
E intermediate distance running	14	2
F long distance running	20	1
G team games	12	3
H 4-a-side practice game	20	2
I full-scale practice game	20	1

Table 3.9

(i) Use the first-fit decreasing algorithm to allocate activities to each of the three training sessions.

(ii) The solution given by the first-fit decreasing algorithm is not satisfactory since it leads to repeated activities in the same session. The first-fit decreasing algorithm is modified so that the next activity is placed in the first available session only if it will fit and if the same activity has not already been placed in that session. Apply this modified algorithm until it fails.

(iii) Prove that it is not possible to fit the activities into the three sessions so that no session contains a repeated activity.

⑥ Eleven boxes are to be packed into crates each of which has a weight limit of 100 kg. There are three boxes of weight 50 kg, three of weight 40 kg, three of weight 30 kg and two of weight 20 kg.

(i) Apply the first-fit decreasing algorithm and state the number of crates used.

(ii) Show that there is a solution using fewer crates. [MEI]

4 Sorting

The first step in the first-fit decreasing algorithm, considered above, involves sorting the list of weights into decreasing order (largest to smallest). Sorting is used to put a list of names into alphabetical order or to rank a list of universities on their 'student satisfaction' scores. Usually such tasks are done using a computer, but how does a computer sort a list?

Sorting is an everyday activity in which the efficiency of the algorithm used is important. There are many popular algorithms for sorting a list of numbers into ascending or descending order.

The sorting algorithms that will be used here are **bubble sort** and **shuttle sort**. You will also meet **quick sort** in the next section.

These algorithms are designed for use by a computer and their efficiency depends on the length of the list and how muddled the items on the list are. A human may well sort a relatively short list more quickly.

Bubble sort algorithm

The bubble sort is so named because numbers which are below their correct positions tend to move up to their proper places, like bubbles in a glass of sparkling drink. On the first pass, the first number in the list is compared with the second and whichever is smaller assumes the first position. The second number is then compared with the third and the smaller is placed in the second position, and so on through the list. At the end of the first pass the largest number will have been left behind in the bottom position.

For the second pass the process is repeated but excluding the last number, and on the third pass the last two numbers are excluded. The list is repeatedly processed in this way until no swaps take place in a pass. The list is then sorted.

Example 3.6

Use the bubble sort to sort the list 7, 5, 2, 4, 10, 1, 6, 3 into ascending order, starting at the left-hand end.

Solution

On the first pass compare 5 and 7, and swap them; then 7 and 2, and swap them; then 7 and 4, and swap them; then 7 and 10, and do not swap them; then 10 and 1, and swap them; then 10 and 6 and swap them; then finally 10 and 3 and swap them. This pass is shown in detail in Figure 3.5. Note that the last number is now in its correct position.

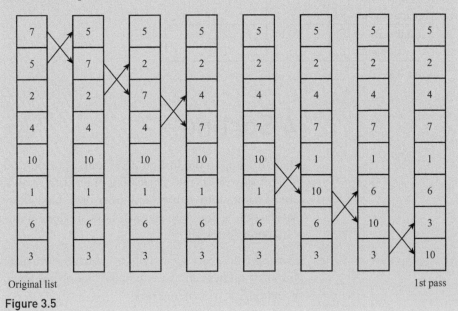

Original list   1st pass

Figure 3.5

ACTIVITY 3.4

The results after making the second and subsequent passes are shown in Figure 3.6. Work through the process to check that you get these results.

2nd pass	3rd pass	4th pass	5th pass	6th pass
2	2	2	1	1
4	4	1	2	2
5	1	4	3	3
1	5	3	4	4
6	3	5	5	5
3	6	6	6	6
7	7	7	7	7
10	10	10	10	10

Figure 3.6

The algorithm for the bubble sort for a list of length 8 can be written in computer pseudo-code like this:

repeat with $i = 1$ to 7

    [repeat with $j = 1$ to $(8 - i)$

    if $L(j) > L(j + 1)$ swap $L(j)$ and $L(j + 1)$]

if no swaps end repeat

The number of comparisons made in a bubble sort for a list of length 8 will be 7 on the first pass, 6 on the second pass, etc. If the maximum number of passes is needed, the total number of comparisons will be $7 + 6 + 5 + 4 + 3 + 2 + 1 = 28$. The number of swaps on the first pass will be anything up to 7; on the second up to 6, etc. So the maximum possible number of swaps will also be: $7 + 6 + 5 + 4 + 3 + 2 + 1 = 28$.

Generalising to a list of size n you can see that the formula will become

$$(n - 1) + (n - 2) + \cdots + 3 + 2 + 1 = \frac{1}{2} n(n - 1),$$

showing that the bubble sort algorithm has quadratic order.

Discussion point

→ Which gives a better measure of the running time for a sorting algorithm – the number of comparisons or the number of swaps?

The shuttle sort algorithm

Start at the left-hand end or top of the list.

1st pass: Compare the first two numbers and swap if necessary to place in ascending order.

2nd pass: Compare the second and third numbers and swap if necessary, then, if a swap occured, compare the first and second numbers and swap if necessary otherwise go to the next pass.

3rd pass: Compare the third and fourth numbers and swap if necessary, then compare the second and third numbers and swap if necessary, and compare the first and second numbers and swap if necessary. If no swap occurs at this point, then go straight to the next pass.

And so on. The sequence is shown in Figure 3.7.

Note

The shuttle sort algorithm also has quadratic complexity. In the worst case scenario: on the first pass there is 1 comparison; on the second pass there are 2 comparisons; on the third pass there are 3 comparisons; on the final $(n-1)$th pass there are $n-1$ comparisons.

So the total number of comparisons is

$1 + 2 + \dots + n - 1$

$= \dfrac{1}{2}(n-1)n.$

n^2 is the dominant term, so shuttle sort has quadratic complexity.

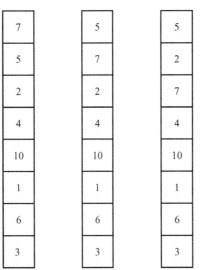

Original list   1st pass   2nd pass   3rd pass

Figure 3.7

Exercise 3.4

① Sort the list 6 5 9 4 5 2 using:

(i) the bubble sort algorithm

(ii) the shuttle sort algorithm.

(iii) which sort required fewer comparisons?

(iv) which sort is more efficient?

② (i) Determine the number of potential swaps when using the shuttle sort for a list of length:

(a) 6

(b) 7

(c) n.

(ii) Explain why the shuttle sort algorithm has quadratic order.

③ It takes 0.02 seconds to sort a list of length 10 using the bubble sort algorithm. How long does it take to sort a list of length 30 using the bubble sort algorithm?

④ At the end of pre-season training, a netball coach has allocated points to ten players. She will choose the seven highest scoring players for her team for the first match.

Player	A	B	C	D	E	F	G	H	I	J
Points	81	92	76	43	82	45	51	93	71	62

Table 3.10

(i) Count the number of comparisons she would have to make if she were to:

1 check along the points table from left to right, comparing the first number with the second, then the larger with the third, and so on to find the player with the highest score

2 choose that player for the team and delete the entry from the points table

3 repeat the process on the reduced points table until the team is chosen.

(ii) Instead she executes a bubble sort on the numbers in the list, starting from the left and with smaller numbers moving to the right. She does this only until she can be sure that the three lowest scores are in the three right-most positions. Show the steps of this sort, and state the number of comparisons that are made. [MEI]

⑤ (i) The list {3, 1, 2} consists of the first three natural numbers in a particular order. This is known as a *permutation* of the numbers {1, 2, 3}. There are six permutations of the numbers {1, 2, 3}. Say how many comparisons each of shuttle sort and bubble sort would need to make to sort them.

(ii) (a) Find a permutation of the set {1, 2, 3, 4, 5, 6, 7} which leads to the largest number of comparisons when sorted by shuttle sort.

Give the number of comparisons.

(b) Find a permutation which leads to the largest number of comparisons when sorted by bubble sort.

Give the number of comparisons.

(c) Without doing a sort, suggest the largest number of comparisons needed to sort a list of length 10 by either shuttle sort or bubble sort. [MEI adapted]

5 More on algorithms

Quick sort

> At this stage the pivot is guaranteed to be in its correct position in the final list and can be marked in some way to indicate this. The pivot splits the list into two sublists: one containing the values that are less than or equal to the pivot (excluding the pivot itself) and the other containing the values that are greater than the pivot. It is possible that one of these sublists may be empty.

To sort a list of numbers into ascending (increasing) order:

1. The first value in the list is the pivot.

   Excluding the pivot, pass along the list and write down each value that is less than or equal to the pivot value, then write the pivot value and then write down the values that are greater than the pivot.

   This concludes the first pass.

2. Repeat Step 1 on each sublist. If a sublist contains just one value this becomes a pivot and is marked as being in its correct position in the final list. This concludes the next pass.

3. Continue in this way until every value is marked as being in its correct position in the final list.

Example 3.7

Use quick sort to sort this list into ascending order:

7   5   2   4   10   1   6   3

> **Note**
>
> Active pivots are boxed and used pivots are underlined.

> **Discussion point**
>
> → How would you adapt quick sort so that the pivot is still the first value in the list (or sublist) but the sort is into descending (decreasing) order?

Solution

Original list	7	5	2	4	10	1	6	3
After 1st pass:	5	2	4	1	6	3	[7]	10
After 2nd pass:	2	4	1	3	[5]	6	7	[10]
After 3rd pass:	1	[2]	4	3	5	[6]	7	10
After 4th pass:	[1]	2	3	[4]	5	6	7	10
After 5th pass:	1	2	[3]	4	5	6	7	10
Sorted list:	1	2	3	4	5	6	7	10

For a small example like this, writing down the sorted list is easy. The example is used to illustrate how quick sort works when it is applied to a much longer list.

You met shuttle sort and bubble sort in the previous section and there algorithms can be compared with each other and with quick sort. For example, you may be asked to count the number of comparisons (or comparisons and swaps) to compare the efficiency of two algorithms being used to sort a particular list.

The worst case for quick sort, in terms of comparisons, is when the pivot at each pass is the smallest or largest value in the sublist (so one of the new sublists is empty). This would be the case when the original list is already sorted (or sorted but in reverse). In the worst case, quick sort has quadratic complexity, $O(n^2)$.

Note

Sometimes the list to be sorted is written vertically; then the two sublists will be above and below the pivot, instead of to the left and to the right of the pivot.

Extending packing methods

The bin packing algorithms you have met so far apply to a one-dimensional situation where you only need to consider packing in one direction.

There are also algorithms for packing in two dimensions.

Example 3.8

Here is an algorithm for placing a number of rectangular tiles into a large rectangular grid without overlap:

1   Take the first tile.

2   Starting from the top left, choose the first vacant square of the large rectangle, searching for it along the top row, then the second row, etc.

3   If possible, place the tile 'lengthways on' (i.e. ▭), so that its top-left corner corresponds to the top-left corner of the chosen grid square. If it can be placed then go to instruction 6.

4   If the tile cannot be placed 'lengthways on' then try it 'end-on' (i.e. ▯) so that its top-left corner corresponds to the top-left corner of the chosen grid square. If it can be placed then go to instruction 6.

5   Choose the next vacant square to the right if there is one and go to instruction 3. If there are no more vacant squares in the current row then choose the left-most vacant square in the next row down and go to instruction 3.

If there are no more rows then go to instruction 7.

6   Take the next unplaced tile from the list and go to instruction 5. If there are no tiles left to place go to instruction 8.

7   Report that the task has not been completed and stop.

8   Report that the task has been completed and stop.

Apply the algorithm to attempt to fit the following six tiles into an 8×4 rectangle.

Tile number	Length	Breadth
1	4	1
2	3	2
3	5	2
4	4	1
5	2	1
6	3	2

Figure 3.8

Draw the positions of the tiles on a copy of the diagram as they are placed.

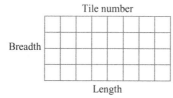

Tile number

Breadth

Length

Figure 3.9

Solution

1   The first tile measures $4 \times 1 \to 2$

2   Select top left corner $\to 3$

3   It is placed lengthways from the top left corner $\to 6$

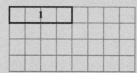

Figure 3.10

6   The second tile measures $3 \times 2 \to 5$

5   Select the fifth square along the top row $\to 3$

3   It is placed lengthways from the fifth square $\to 6$

Figure 3.11

6   The third tile measures $5 \times 2 \to 5$

5   Select the last square in the top row $\to 3$

3   It cannot be placed $\to 4$

4   It cannot be placed $\to 5$

5   There is no further square in the top row so choose the first square in the second row $\to 3$

3   It cannot be placed $\to 4$

4   It cannot be placed $\to 5$

5   Select the second square in the second row $\to 3$

3   It cannot be placed $\to 4$

4   It cannot be placed $\to 5$

5   Select the third square in the second row $\to 3$

3   It cannot be placed $\to 4$

4   It cannot be placed $\to 5$

5   Select the fourth square in the second row $\to 3$

3   It cannot be placed $\to 4$

4	It cannot be placed → 5
5	Select the eighth square in the second row → 3
3	It cannot be placed → 4
4	It cannot be placed → 5
5	Select the first square in the third row → 3
3	It is placed lengthways from the first square → 6

Figure 3.12

6	The fourth tile measures 4 × 1 → 5
5	Select the sixth square in the third row → 3
3	It cannot be placed → 4
4	It cannot be placed → 5
5	Select the seventh square in the third row → 3
3	It cannot be placed → 4
4	It cannot be placed → 5
5	Select the eighth square in the third row → 3
3	It cannot be placed → 4
4	It cannot be placed → 5
5	Select the sixth square in the fourth row → 3
3	It cannot be placed → 4
4	It cannot be placed → 5
5	Select the seventh square in the fourth row → 3
3	It cannot be placed → 4
4	It cannot be placed → 5
5	Select the eighth square in the fourth row → 3
3	It cannot be placed → 4
4	It cannot be placed → 5
7	The task has not been completed. Stop.

The algorithm leaves three of the tiles unplaced.

ACTIVITY 3.5

Attempt to fit the tiles in Example 3.8 into the 8 × 4 rectangle using a heuristic method.

Knapsack problems

A variation on a simple packing problem is where the items have a value as well as a size. The sizes provide a constraint on the packing, as with the other methods you have met in this chapter. The values introduce a need to optimise the packing to maximise the value, perhaps the profit or usefulness of what is packed.

Example 3.9

Agnes is travelling by train to sell some items at a craft fair. She can carry up to 25 kg in her suitcase. Looking at the items she wishes to take she realises that she cannot take all of them. She lists them, together with their weight and the profit she would make by selling them.

Item	Weight (kg)	Profit (£)
1	2	5
2	3	6
3	5	4
4	8	8
5	13	10
6	14	9

Table 3.11

(i) List the items she should pack to maximise her profit.

(ii) Describe another aspect that she may prefer to take into account.

Solution

(i) Looking at the highest value items, she cannot take both, as $13 + 14 > 25$.

Item 6 weighs more than item 5, but has less value, so item 5 is chosen in preference.

The other four items weigh 18 kg, but $13 + 18 > 25$ by 6 kg.

Removing the 8 kg item reduces the total value by less than any combination of two of the three smaller items.

$13 + 5 + 3 + 2 = 23$ kg with a profit of £25 is the best solution.

A more efficient method might be to calculate the 'density' of the profit and order the items according to that.

Item	Weight (kg)	Profit (£)	Density
1	2	5	2.5
2	3	6	2
3	5	4	0.8
4	8	8	1
5	13	10	0.77
6	14	9	0.64

Table 3.12

The ordered list becomes 2, 3, 8, 5, 13, 14

Taking the items in order until there is no room for the next one would yield:

$2 + 3 + 8 + 5 = 18$

This leaves room for another 7 kg, and so there is reason to suspect that a better solution may be found.

This initial packing then allows swapping items to improve the solution.

➔

ACTIVITY 3.6

(a) For each of the permutations of the letters A, B, C, D determine the number of comparisons required when using the quick sort algorithm to place them in alphabetical order.

(b) How many comparisons are there in the best case?

(c) What is the average number of comparisons?

(d) How many comparisons are there in a typical case?

(e) Is this consistent with the hierarchy of orders above?

Swapping either 13 or 14 for 5 (the lowest density item in the suitcase) will add 8 kg or 9 kg respectively to the weight, and that is more than the 7 kg 'space' available.

Swapping 8 for either the 13 or 14 is within the weight constraint and the larger profit on 13 makes it a better choice.

Swapping out either of the smaller items would not allow the weight constraint to be satisfied.

This gives the same solution as before.

(ii) The total value of the items rather than the profit may be more helpful to cash flow; the size of the items may affect whether they can be carried in the suitcase.

Orders of algorithms

You have seen that the order of an algorithm gives information about the way the run-time increases as the size of the problem increases. Some sorting algorithms are of O(n^2) and so have quadratic complexity; doubling the number of items to be sorted approximately quadruples the run-time. The worst-case scenario for quick sort is O(n^2) but the average is O$(n \log n)$. It is useful to use the hierarchy of orders so that you may choose a more efficient algorithm if your problem is large.

$$O(1) \subset O(\log n) \subset O(n) \subset O(n \log n) \subset O(n^2) \subset O(n^3) \subset \cdots \subset O(a^n) \subset O(n!)$$

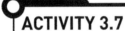

ACTIVITY 3.7

Repeat Activity 3.6 for the permutations of the letters A, B, C, D, E.

Exercise 3.5

① (i) Sort the list 6   5   9   4   5   2 into ascending order using a quick sort.

(ii) How many comparisons are made?

(iii) Explain why, in the worst-case scenario, quick sort has quadratic complexity.

② Diana is choosing between three algorithms to solve a problem. The order of algorithm A is O(n^2), algorithm B is O$(\log n)$ and algorithm C is O(2^n). Advise her which would be best.

③ Change the algorithm in Example 3.8, by replacing instruction 6 with

6 Take the next unplaced tile from the list and go to instruction 2. If there are no further tiles left to place, go to instruction 8.

(i) Use the set of tiles in Example 3.8 with the 8 × 4 rectangle and apply the modified algorithm to them.

(ii) Explain why it leaves one tile unplaced.

(iii) Describe a further improvement to the modified algorithm in (i) that means all of the tiles are placed.

④ A hiker going on a camping trip can carry a weight of up to 45 pounds in his knapsack. There are five items that he had hoped to take, but to take them all would exceed his weight allowance. He has assigned a value to each item. The items and their weights and values are as follows.

Item	Weight	Value
1	13	7
2	12	9
3	15	30
4	16	16
5	18	27

Table 3.13

What items should he pack so that their total value is a maximum, subject to the weight restriction?

⑤ Use a quick sort in this question.

(i) Sort 5 6 7 4 1 2 3 into ascending order. How many comparisons were made?

(ii) Find the number of comparisons needed to sort 7 6 5 4 3 2 1 into ascending order.

(iii) Determine the least number of comparisons required to sort a permutation of the numbers 1 to 7 into ascending order.

⑥ A cuboidal container measures 8 m by 3 m by 2 m. There are seven boxes, as follows:

A, B, C	3 m × 2 m × 1 m
D	4 m × 2 m × 2 m
E	2 m × 2 m × 2 m
F, G	3 m × 1 m × 1 m

Table 3.14

Is it possible to fit all of the boxes into the container? Justify your answer, including the heuristic that you used.

⑦ Use a quick sort, starting at the left-hand end, in this question.

(i) (a) Determine the number of comparisons required to sort 4 3 2 1 into ascending order.

(b) Determine the number of comparisons required to sort 2 4 1 3 into ascending order.

(ii) (a) Determine the number of comparisons required to sort 8 7 6 5 4 3 2 1 into ascending order.

(b) Determine the number of comparisons required to sort 4 1 3 8 7 2 5 6 into ascending order.

(iii) Comment on your results in parts (i) and (ii).

LEARNING OUTCOMES

Now you have finished this chapter you should be able to

➤ understand that an algorithm has an input and an output, is deterministic and finite

➤ appreciate why an algorithmic approach to problem solving is generally preferable to ad hoc methods, and understand the limitations of algorithmic methods

➤ be able to trace through an algorithm and interpret what the algorithm has achieved

➤ see that algorithms may be presented as flow diagrams, listed in words, or written in simple pseudo-code

➤ use the order of an algorithm to calculate an approximate run-time for a large problem by scaling up a given run-time

➤ compare the efficiency of two algorithms that achieve the same end result by considering a given aspect of the run-time in a specific case

➤ calculate worst case time complexity, the 'maximum run-time' $T(n)$, as a function of the size of a problem, by considering the worst case for a specific problem

➤ be familiar with $O(n^k)$, where n is a measure of the size of the problems and $k = 0, 1, 2, 3$ or 4

➤ be familiar with the next-fit, first-fit, first-fit decreasing and full bin methods for one-dimensional packing problems

➤ sort a list using bubble sort and using shuttle sort.

➤ calculate the run-time as a function of the size of a problem by considering the best case, the worst case or a typical case

➤ be familiar with:

➤ $O(n^k)$ for $k \in \mathbb{R}$,

➤ $O(a^n)$ for $a > 0$,

➤ $O(\log n)$

where n is a measure of the size of the problem

➤ sort a list using quick sort

➤ extend your knowledge of packing methods.

KEY POINTS

1 An algorithm is a finite sequence of operations for carrying out a procedure or solving a problem that may be communicated in ordinary language, in a flowchart or in pseudo-code.

2 For large n, an algorithm with $O(n^a)$ complexity is more efficient than an algorithm with $O(n^b)$ complexity, when $a < b$.

3 If an algorithm has $O(n^r)$ complexity, then scaling the problem size by a factor k will scale the run-time by a factor of (approximately) k^r.

4 **Next-fit algorithm:** Take the boxes in the order listed and pack each box in the first bin that has enough space for it, starting from the current bin.

5 **First-fit algorithm:** Take the boxes in the order listed and pack each box in the first bin that has enough space for it (each time starting with the first bin).

6 **First-fit decreasing algorithm**: Reorder the boxes from the largest to the smallest, then apply the first-fit method to this list.

7 **Full-bin strategy:** Look for combinations of boxes to fill bins. Pack these boxes. Put the remaining boxes together in combinations that result in bins that are as nearly full as possible.

8 A heuristic is a method that finds a solution efficiently, but with no guarantee that the solution is optimal. Heuristics are important when classic methods fail; for example when the only way to guarantee finding the optimal solution is a complete enumeration.

9 The efficiency of different packing strategies can be compared by counting the number of comparisons needed in the worst case for a list of n items. In the worst case, first-fit and first-fit decreasing both have quadratic complexity, $O(n^2)$.

10 **Bubble sort:** On the first pass, the first number in the list is compared with the second and the smaller assumes the first position. The second and third numbers are then compared with the smaller assuming the second position and so on to the end of the list. The largest number is now at the end of the list.

Repeat with all but the last number. Continue until there are no swaps in a complete pass.

11 **Shuttle sort**

Start at the left-hand end or top of the list and go to the next pass as soon as no swap is necessary.

1st pass: compare the first two numbers and swap if necessary to place in ascending order.

2nd pass: compare the second and third numbers and swap if necessary, then compare the first and second numbers and swap if necessary.

3rd pass: compare the third and fourth numbers and swap if necessary, then compare the second and third numbers and swap if necessary, and compare the first and second numbers and swap if necessary.

And so on.

12 **Quick sort**

The first value in the list is the pivot. Excluding the pivot, pass along the list and write down each value that is less than or equal to the pivot value, then write the pivot value and then write down the values that are greater than the pivot. The pivot splits the list into two sub-lists.

Repeat the process on each sub-list and continue in this way until every value has been a pivot.

13 In the worst case, quick sort has quadratic complexity, $O(n^2)$.

4

'Tis true; there's magic in the web of it.

Shakespeare, Othello

Network algorithms

→ What is the shortest route that visits A, B, C and D in the network below and returns to its starting point?

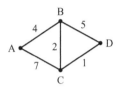

1 The language of networks

A network is a **weighted** graph – i.e. a graph for which there is a number (**weight**) associated with each arc.

Figure 4.1 shows a graph from Chapter 2 with weights attached.

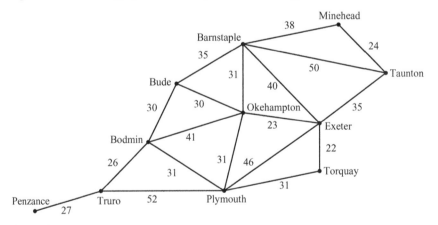

Figure 4.1

In this case, the weights are the distances in miles between the various towns and cities. In other situations the weights could be times or costs, for example.

A **table of weights** (or distance matrix, if applicable) can be used to represent the network in a form that is convenient for a computer. The network in Figure 4.1, which has no loops, multiple arcs or directed arcs, can be represented by Table 4.1.

	Pen	Tru	Ply	Tor	Exe	Tau	Min	Bar	Oke	Bud	Bod
Pen	–	27	–	–	–	–	–	–	–	–	–
Tru	27	–	52	–	–	–	–	–	–	–	26
Ply	–	52	–	31	46	–	–	–	31	–	31
Tor	–	–	31	–	22	–	–	–	–	–	–
Exe	–	–	46	22	–	35	–	40	23	–	–
Tau	–	–	–	–	35	–	24	50	–	–	–
Min	–	–	–	–	–	24	–	38	–	–	–
Bar	–	–	–	–	40	50	38	–	31	35	–
Oke	–	–	31	–	23	–	–	31	–	30	41
Bud	–	–	–	–	–	–	–	35	30	–	30
Bod	–	26	31	–	–	–	–	–	41	30	–

Table 4.1

2 The minimum connector problem

Suppose that a cable company wishes to join up the towns and cities in Figure 4.1, using the shortest possible length of cable. This can be done by creating a spanning tree with the minimum weight.

There are several ways of doing this. One method is simply to remove arcs in order of decreasing weight, ensuring that the network remains connected.

Example 4.1

Find a minimum spanning tree for the network in Figure 4.1.

Solution

The arcs can be removed in the following order (arcs of the same weight may be chosen arbitrarily):

Tru–Ply 52

Bar–Tau 50

Ply–Exe 46

Bod–Oke 41

Bar–Exe 40

Bar–Min 38

Bud–Bar 35

(Exe–Tau 35 can't be removed, as the network would no longer be connected)

Ply–Tor 31

Bod–Ply 31

(Ply–Oke 31 can't be removed)

→

The resulting spanning tree is shown in Figure 4.2. It has a total weight of 279.

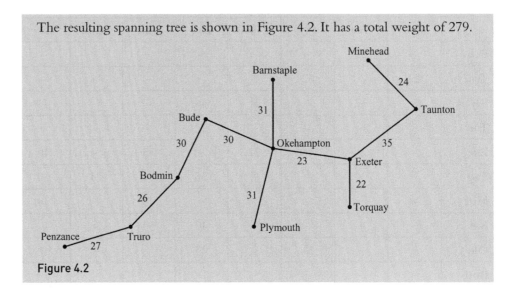

Figure 4.2

Some methods may be easier for a computer to apply. One such method is **Prim's algorithm**.

(i)   Start with any node.

(ii) Add the arc leading to the nearest node.

(iii) Add the arc leading (from any of the nodes collected so far) to the nearest new node, and repeat.

(iv) Stop once all nodes have been connected.

If two nodes are the same distance from the nodes connected so far, either node may be chosen.

Example 4.2

Use Prim's algorithm to find a minimum spanning tree for the network in Figure 4.1, starting from Okehampton. Make clear the order in which the arcs are selected.

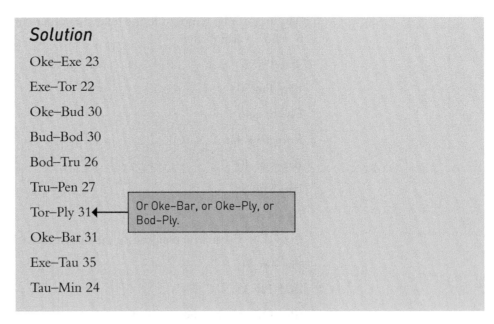

Solution

Oke–Exe 23

Exe–Tor 22

Oke–Bud 30

Bud–Bod 30

Bod–Tru 26

Tru–Pen 27

Tor–Ply 31 ◄—— Or Oke-Bar, or Oke-Ply, or Bod-Ply.

Oke–Bar 31

Exe–Tau 35

Tau–Min 24

The resulting spanning tree is shown in Figure 4.3.

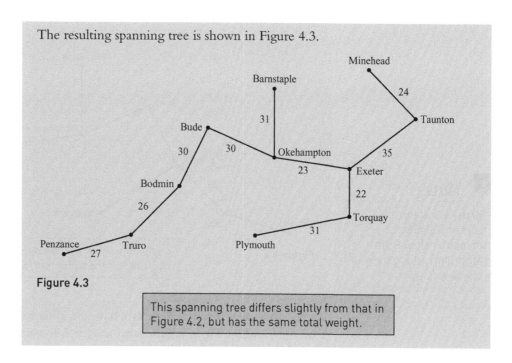

Figure 4.3

This spanning tree differs slightly from that in Figure 4.2, but has the same total weight.

Another well-known method is **Kruskal's algorithm**:

(i)   Start with the shortest arc.
(ii) Choose the next shortest arc, provided it doesn't create a cycle, and repeat.
      (If two arcs are of equal length, then either may be chosen.)
The order in which arcs are added should be made clear.

Example 4.3

Use Kruskal's algorithm to find a minimum spanning tree for the network in Figure 4.1, starting from Okehampton. Make clear the order in which the arcs are selected.

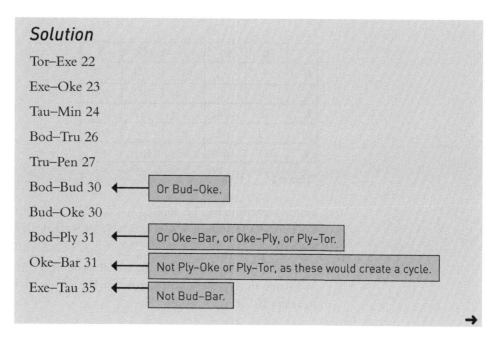

Solution

Tor–Exe 22

Exe–Oke 23

Tau–Min 24

Bod–Tru 26

Tru–Pen 27

Bod–Bud 30 ← Or Bud–Oke.

Bud–Oke 30

Bod–Ply 31 ← Or Oke-Bar, or Oke-Ply, or Ply-Tor.

Oke–Bar 31 ← Not Ply-Oke or Ply-Tor, as these would create a cycle.

Exe–Tau 35 ← Not Bud–Bar.

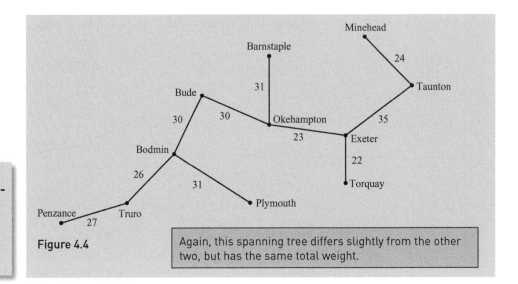

Figure 4.4

Again, this spanning tree differs slightly from the other two, but has the same total weight.

As you have seen, if some of the arcs have the same weight then there may be more than one solution. However, the total weight will always be the same.

Matrix representation for Prim's algorithm

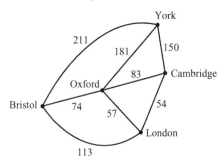

Figure 4.5

	B	**C**	**L**	**O**	**Y**
B	–	–	115	74	211
C	–	–	54	83	150
L	115	54	–	57	–
O	74	83	57	–	181
Y	211	150	–	181	–

Table 4.2

Prim's algorithm can be applied to the table of weights for a network. Referring to the network in Figure 4.5, and its table of weights in Table 4.2, the procedure is as follows:

Choosing B (for example) as the initial node, you can write (1) above B's column and shade in (or cross out) the row for B (to indicate that it is no longer necessary to select B). See Table 4.3.

	(1)				
	B	**C**	**L**	**O**	**Y**
B	–	–	115	74	211
C	–	–	54	83	150
L	115	54	–	57	–
O	74	83	57	–	181
Y	211	150	–	181	–

Table 4.3

Then find the smallest weight in column B: 74 for O (indicating that O is the nearest node to B). Write (2) above O's column, and place a bracket (or circle) around the 74, for future reference, as well as shading in the row for O (see Table 4.4).

	(1)			(2)	
	B	**C**	**L**	**O**	**Y**
B	–	–	115	74	211
C	–	–	54	83	150
L	115	54	–	57	–
O	[74]	83	57	–	181
Y	211	150	–	181	–

Table 4.4

Arc BO has now been created. Then find the smallest weight in either of columns B and O (i.e. look for the node that is nearest to one of the nodes selected so far). This is 57 for L in column O. This is recorded in the same way as before. See Table 4.5.

	(1)		(3)	(2)	
	B	**C**	**L**	**O**	**Y**
B	–	–	115	74	211
C	–	–	54	83	150
L	115	54	–	[57]	–
O	[74]	83	57	–	181
Y	211	150	–	181	–

Table 4.5

Tree BOL has now been created. The process then continues until all of the nodes have been included.

Complexity

The run-time for Kruskal's algorithm will be approximately proportional to the number of comparisons made when selecting the arcs in order of increasing size. (This ignores the work involved in checking that no cycles have been created.) Considering a complete graph with n nodes (as the worst case), there will be $\frac{1}{2}(n-1)n$ arcs in total (as seen in Chapter 2), and hence $\frac{1}{2}(n-1)n - 1$ comparisons

Note

$$\frac{1}{2}(n-2)(n-1)\,n$$
$$= \frac{1}{2}n^3 - \frac{3}{2n^2} + n$$

For large n, the size of $\frac{1}{2}n^3$ far exceeds the size of the other terms and so is the dominant term. The coefficient $\frac{1}{2}$ is a constant and so n^3 determines the scaling up of the run-time, $T(n)$, for larger problems.

ACTIVITY 4.1

A network consists of 5 nodes and an algorithm takes 1.5 milliseconds to find the minimum spanning tree. Approximately how long will it take to find the minimum spanning tree for a network of 20 nodes?

will be needed when selecting the shortest arc. There will then be $\frac{1}{2}(n-1)n-2$ comparisons for the next shortest arc. Overall there will be

$[\frac{1}{2}(n-1)n-1] + [\frac{1}{2}(n-1)n-2]... + [\frac{1}{2}(n-1)n-(n-1)]$ comparisons, and this equals $\frac{1}{2}(n-1)n(n-1) - (1 + 2 + \cdots + (n-1))$

$$= \frac{1}{2}(n-1)^2 n - \frac{1}{2}(n-1)n$$
$$= \frac{1}{2}(n-1)n([n-1]-1)$$
$$= \frac{1}{2}(n-2)(n-1)n$$

Thus Kruskal's algorithm has **cubic complexity**.

Now consider a complete graph with n nodes in the case of Prim's algorithm. The first step involves selecting the nearest node out of $n-1$ possibilities: $(n-1)-1$ comparisons are required. Having selected two nodes, each of the remaining $n-2$ nodes could be joined to either of these two nodes: $2(n-2)-1$ comparisons are required. Overall there will be $[(n-1)-1] + [2(n-2)-1] + [3(n-3)-1]... + [(n-1)(1)-1]$ comparisons, and this equals

$$\left\{\sum_{r=1}^{n-1} r\,(n-r)\right\} - (n-1)$$
$$= \left\{n\sum_{r=1}^{n-1} r\right\} - \left\{\sum_{r=1}^{n-1} r^2\right\} - (n-1)$$
$$= n \times \left(\frac{1}{2}\right) \times (n-1)n - \frac{1}{6}(n-1)n(2n-1) - (n-1), \text{ using standard summation formulae.}$$

As this expression involves the term $\frac{1}{2}n^3 - \frac{1}{3}n^3 = \frac{1}{6}n^3$, Prim's algorithm also has cubic complexity.

Exercise 4.1

① Apply Prim's algorithm to Figure 4.1, starting at Bodmin.

② Create a table of weights to represent the network in Figure 4.6.

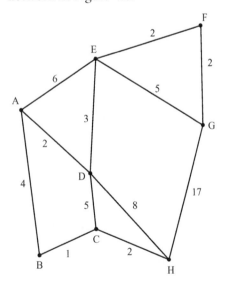

Figure 4.6

③ Draw the network with the table of weights in Table 4.6.

	A	B	C	D	E	F
A	–	10	20	–	16	12
B	10	–	15	–	–	–
C	20	15	–	11	8	–
D	–	–	11	–	7	22
E	16	–	8	7	–	14
F	12	–	–	22	14	–

Table 4.6

④ Apply Prim's algorithm to the network in question 3, starting at A. Show the order in which the steps are carried out, and give the total weight of the minimum spanning tree.

⑤ A computer takes 0.5 seconds to run Prim's algorithm for a network with 15 nodes. Estimate how long it would take for a network with 60 nodes.

⑥ Apply Kruskal's algorithm to the network in question 3. Show the order in which the steps are carried out.

⑦ Find the minimum spanning tree for the following network which is given in tabular form.

	Malvern	Worcester	Hereford	Evesham	Ross	Tewkesbury	Gloucester	Cheltenham
Malvern	–	8	19	–	19	13	20	–
Worcester	8	–	25	16	–	15	–	–
Hereford	19	25	–	–	14	–	28	–
Evesham	–	16	–	–	–	13	–	16
Ross	19	–	14	–	–	24	16	–
Tewkesbury	13	15	–	13	24	–	10	9
Gloucester	20	–	28	–	16	10	–	9
Cheltenham	–	–	–	16	–	9	9	–

Table 4.7

⑧ Find the minimum spanning tree for this network.

	Dorchester	Puddletown	Blandford	Wimborne	Bere Regis	Lytchett Minster	Weymouth	Warmwell	Wareham	Swanage	Poole
Dorchester	–	5	–	–	–	8	5	–	–	–	
Puddletown	5	–	12	–	6	–	–	9	14	–	–
Blandford	–	12	–	7	9	11	–	–	16	–	–
Wimborne	–	–	7	–	8	7	–	–	–	–	7
Bere Regis	–	6	9	8	–	8	19	11	8	–	–
Lytchett Minster	–	–	11	7	8	–	25	–	5	–	6
Weymouth	8	–	–	–	19	25	–	7	–	–	–
Warmwell	5	9	–	–	11	–	7	–	13	–	–
Wareham	–	14	16	–	8	5	–	13	–	10	–
Swanage	–	–	–	–	–	–	–	–	10	–	–
Poole	–	–	–	7	–	6	–	–	–	–	–

Table 4.8

3 Dijkstra's algorithm

Dijkstra's algorithm provides a procedure for determining the shortest path between two particular nodes of a network.

Consider the network in Figure 4.7.

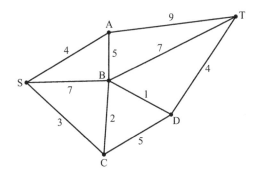

Figure 4.7

Start by recording provisional shortest distances from S to each of A, B and C. These are just the lengths of the arcs SA, SB and SC, and you say that A, B and C have been given **temporary labels** of 4, 7 and 3 respectively.

In the case of B, you can see that the distance of 7 can be improved on, by following the path SCB, and this will be taken into account later on.

For the moment, all that can be stated with certainty is that the shortest distance from S to C is 3, as any other route from S to C would involve passing through A or B, and the distances SA and SB are both greater than the distance SC. So a **permanent label** of 3 is assigned to C.

Figure 4.8 shows a way of carrying out the labelling.

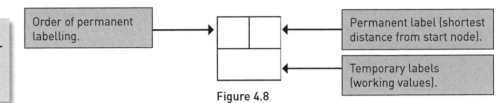

Figure 4.8

For completeness, S can also be given a permanent label of 0.

In the same way that the immediate neighbours of S were given temporary labels, a temporary label is now given to the immediate neighbours of C, based on the direct path from C. For D, this is $3 + 5 = 8$. B's temporary label can also be improved upon: going via C, B can be reached in a distance of $3 + 2 = 5$.

The current position is shown in Figure 4.9.

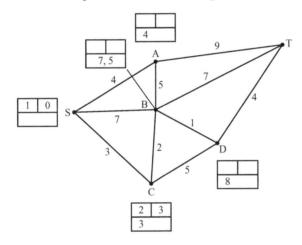

Figure 4.9

You now know that the shortest distance from S to A must be 4, being the smallest of the temporary labels of A, B and D, so A's label is made permanent.

Proof

A proof of this (which may be omitted on a first reading) is as follows:

Suppose that the last arc on the shortest route from S to A is not SA. More generally, suppose that the last arc leading into A does not come from a permanently labelled node. Any node that it does come from must be reached via one of the existing permanently labelled nodes (including S itself). This means that the shortest path must have passed along one of the arcs leading from a permanently labelled node to a temporarily labelled node (as all the neighbours of the permanently labelled nodes have been given temporary labels – unless they already have permanent labels). But no temporary label is smaller than that of A, and so the supposed shortest route has already covered a distance equal to A's temporary label. So it is not possible to improve on A's temporary label, and it can therefore be made permanent.

If more than one node shares the smallest temporary label, it doesn't matter which is made permanent.

The process is now repeated: the immediate neighbours of A are given temporary labels, or their temporary labels are improved on. The position after this step is shown in Figure 4.10.

ACTIVITY 4.2

Complete this procedure. (The solution is given below.)

Note

Dijkstra's algorithm can be applied to networks with directed arcs, but not to networks with negative weights.

Discussion point

→ How could you deal with the situation where there are several possible starting points (say S_1, S_2, ...) and you wish to find the shortest route to a single end point (say T)?

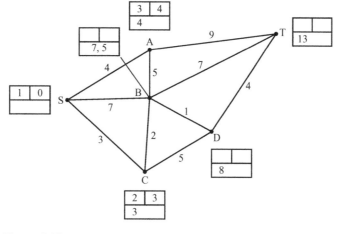

Figure 4.10

B now has the smallest temporary label, which is therefore made permanent.

The final position is shown in Figure 4.11.

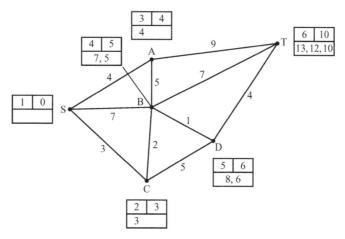

Figure 4.11

Discussion point

→ Why do you work backwards from T?

The shortest distance from S to T is thus 10. Working backwards from T, the shortest route is determined by looking for neighbouring nodes for which the length of the arc between them equals the difference between their permanent labels.

In this example, the last stage has to be DT (since $10 - 6 = 4$, which is the arc length). Then BD = 1 is the next leg, followed by CB = 2, and finally SC = 3. Thus the shortest route is SCBDT, with a total distance of $3 + 2 + 1 + 4 = 10$ (which, as a check, equals the label of T).

Note that, although the aim was to find the shortest route from S to T, the algorithm also finds the shortest route from S to each of the other nodes.

Note

Dijkstra's algorithm can be applied to networks with directed arcs, but not to networks with negative weights.

Complexity

Dijkstra's algorithm can be shown to have quadratic complexity, as follows.

As usual, the worst case of a complete graph is assumed. When r permanent labels have been created, the amount of work involved in updating the temporary labels will be approximately proportional to $n - r$.

So the total amount of work is approximately proportional to

$$(n - 1) + (n - 2) + \cdots + 1 = \tfrac{1}{2}(n - 1)n.$$

ACTIVITY 4.3

A network consists of 20 nodes and an algorithm takes at most 2 milliseconds to find the shortest route from one node to another. What is the new time likely to be if there are 200 nodes?

Exercise 4.2

① Use Dijkstra's algorithm to find a shortest path from S to T for the networks (i), (ii) and (iii) below.

(i)

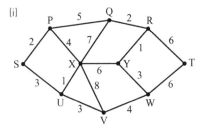

(ii)

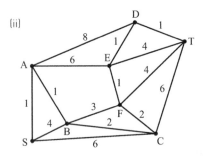

(iii)

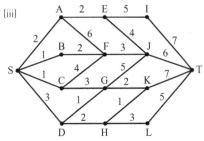

Figure 4.12

② The map shows the main railway lines across the USA and gives the approximate times in hours for the various journeys.

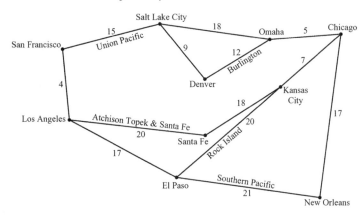

Figure 4.13

(i) Find the quickest route from Los Angeles to Chicago.

(ii) Find the quickest route from New Orleans to Denver.

(iii) If you can travel by road from El Paso to Santa Fe in 5 hours and from Santa Fe to Denver in 5 hours, would you save time on journey (i) or (ii) by using a mix of road and rail? (You should neglect connection times.)

③ The fire department in Westingham has a team fighting a large blaze at one of the town's hotels. They urgently need extra help from one of the neighbouring towns, A, B or C. The estimated times (in minutes) to travel along the various sections of road from A, B and C to Westingham are shown on the network in Figure 4.14. Which town's fire fighters should they call upon and how long will it take them to arrive at the fire?

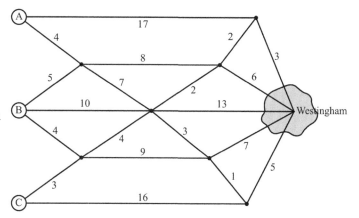

Figure 4.14

4 The route inspection problem

This is also known as the Chinese postman problem.

In Chapter 2, you met the example of the gritting truck that needs to pass along all the arcs exactly once. An Eulerian graph was defined as one where this was possible, and in such a way that the truck would be able to return to its starting point. A semi-Eulerian graph was defined as one where the truck could pass along all of the arcs exactly once, but would not be able to return to its starting point.

The route inspection problem involves first establishing whether the graph (i.e. the network without its weights) is Eulerian, semi-Eulerian or neither. Depending on the type of graph identified, you may then need to make some compromise in order to find the shortest route that covers all of the arcs at least once – returning to the starting point, if required.

You saw in Chapter 2 that the number of odd nodes in a graph has to be even, and that graphs can be divided into the following three categories.

A Those with no odd nodes, which are therefore Eulerian.

B Those with two odd nodes, which are therefore semi-Eulerian.

C Those with four or more odd nodes, which are neither Eulerian nor semi-Eulerian.

In the case of category **A** no compromise has to be made: each of the arcs is covered exactly once, and you end up back at the starting point. The length of the shortest route is just the total of the weights of the network.

Although the existence of such a route is guaranteed, you still need to find it. However, there is usually more than one solution, and it normally isn't difficult to find an example.

In the case of category **B**, if it is permissible for the start and end nodes to be different, then no compromise is necessary. However, if you wish to end up where you started, then you have to convert the network into one that has no odd nodes.

This is done by finding the shortest possible path between the two odd nodes, and duplicating that path (so that there are multiple arcs between some pairs of nodes). In this way, the two odd nodes are made even, and any nodes along the duplicated path have their degree increased by two (and so will still be even).

For the network shown in Figure 4.15, find the shortest route such that all arcs are covered at least once and which returns to the starting point.

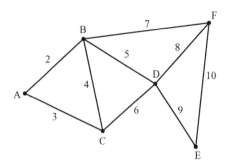

Figure 4.15

Solution

Here there are two odd nodes: C and F.

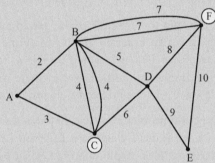

Figure 4.16

It is not necessary for BF to follow CB in the shortest route: it is sufficient that the arcs CB and BF are repeated at some point.

As the shortest route between them is CBF (of weight $4 + 7 = 11$), this path is repeated in the network, as shown in Figure 4.16.

The total length of the original arcs is 54, and the effect of adding on the repeated arcs is to give a value of $54 + 4 + 7 = 65$ for the shortest route such that all arcs are covered at least once and that returns to the starting point.

ACTIVITY 4.4

For the network given in Figure 4.16, find a route that starts and ends at A.

In the case of category **C** where there are more than two odd nodes, there are several possibilities.

(i) You must return to the starting point.
(ii) You can start and finish where you like.
(iii) You must start or finish at a specified node.

For (i), the approach is similar to that for category **B**: you decide how to pair up the odd nodes in such a way that the total of the shortest distances between the paired nodes is minimised.

To do this, you first find the shortest path between each possible pair of odd nodes (AB, AC, AD, BC, BD, CD – where the four odd nodes are A, B, C and D).

Then you establish all the possible ways of pairing up the odd nodes (in the case of four nodes, there will be three possibilities: AB & CD, AC & BD and AD & BC).

You then choose the combination of pairings that gives the shortest total path (e.g. AC & BD, if AC + BD is the smallest possible total).

This total is the additional distance that has to be added to the original total of all the arcs, to give the length of the shortest route.

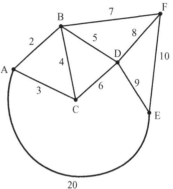

Example 4.5

For the network given in Figure 4.17, find the shortest route that allows all arcs to be covered at least once and which returns to the starting point.

Figure 4.17

ACTIVITY 4.5

For the network given in Figure 4.17, find a route that starts and ends at A.

ACTIVITY 4.6

For Figure 4.17, what would be the best strategy if you had to start at A and finish at B (an example of category C(iii))?

ACTIVITY 4.7

Find the shortest route if you have to start at A and finish at B.

Solution

There are now four odd nodes: A, C, E and F.

The possible pairings, together with the shortest distances associated with them, are (by inspection) as follows.

AC 3

AE 16 (ABDE)

AF 9 (ABF)

CE 15 (CDE)

CF 11 (CBF)

EF 10

The possible ways of pairing up these nodes, together with the total distances in each case, are

(AC) (EF)   3 + 10 = 13

(AE) (CF)   16 + 11 = 27

(AF) (CE)   9 + 15 = 24.

The combination that gives the shortest total distance is thus (AC) (EF).

Since the total of the original arcs is 74, the effect of adding the repeated arcs is to give a value of 74 + 13 = 87 for the shortest route that allows all arcs to be covered at least once and returns to the starting point.

Discussion point

➔ In the case of Figure 4.17, what would be the best strategy for category C(ii), where you can choose the start and end nodes?

Example 4.6

Arnold, who is a railway enthusiast, wishes to travel along each stretch of railway linking the cities A–H of a particular country, as shown in Figure 4.18, with the times (in hours) for each stretch. The total time for all the stretches is 57 hours.

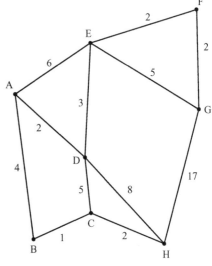

Figure 4.18

(i) Initially he plans to set out from A and return to A. Find a route that covers each stretch of railway at least once, in the shortest possible time, and find the time taken.

(ii) There is a change of plan, and now Arnold wishes to start at A and finish at H (still covering each stretch at least once). Find the new time taken for the quickest route.

(iii) Arnold's wife wants the time reduced. If he has a free choice as to the starting and finishing cities, which should he choose, and what will the new time be?

Solution

(i) First, the degrees of each node are established.

A 3 B 2 C 3 D 4 E 4 F 2 G 3 H 3

Thus there are 4 nodes of odd degree: A, C, G and H.

In order to create an Eulerian graph (one where you can return to the starting position, having travelled along each arc exactly once), the nodes of odd degree need to be converted to even degree, by adding in extra arcs (which will be repeats of some of the existing arcs).

For example, you might join up A and C, and then G and H.

The other possibilities are AG and CH, and AH and CG.

When joining up these nodes, you need to use paths between them that have the smallest total weight.

These total weights are as follows.

AC and GH: 5[ABC] + 14[GFEDCH] = 19

AG and CH: 9[ADEFG] + 2[CH] = 11

AH and CG: 7[ABCH] + 12[CDEFG] = 19

So the best option is AG and CH, which involves repeating the arcs AD, DE, EF, FG and CH (as shown in Figure 4.19).

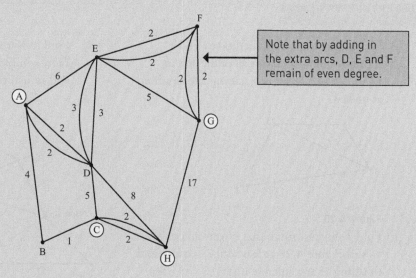

Note that by adding in the extra arcs, D, E and F remain of even degree.

Figure 4.19

One possibility is ABCDADEDHCHGEFGFEA

The time taken is then 57 + 11 = 68 hours.

(ii)   Starting at A and finishing at H means that nodes A and H can remain of odd degree, so that only C and G need to be joined up.

This gives rise to an additional 12 hours (adding CDEFG), and hence the total time is now 57 + 12 = 69 hours.

(iii)   Of the available options for joining up nodes of odd degree, CH has the smallest weight (of 2). Hence the best option is to start at A and finish at G, or vice versa, so that CH is travelled along twice.

This gives a total time of 57 + 2 = 59 hours.

Number of pairings

The number of possible pairings increases dramatically as the number of odd nodes increases. Suppose that there are 8 odd nodes: A–H. One possible set of pairings is (AB)(CD)(EF)(GH).

There are 7 odd nodes that can be paired with node A. Then, for each of these pairs, there will be 5 odd nodes that can be paired with any one of the remaining odd nodes (C, say). Once the next two odd nodes have been paired up, there will be 3 odd nodes that can be paired with any one of the remaining odd nodes (E, say), and so on.

So the number of possible pairings for 8 odd nodes is $7 \times 5 \times 3 \times 1 = 105$.

For $2n$ odd nodes the number of possible pairings will be

$(2n - 1)(2n - 3) \ldots \times 5 \times 3$

ACTIVITY 4.8

Show that $(2n - 1)(2n - 3) \ldots \times 5 \times 3$ can be written as $\frac{(2n)!}{n!2^n}$ and determine the number of possible pairings when there are 20 odd nodes.

Exercise 4.3

① (i) Use a route inspection method to find a shortest route that covers all of the arcs in Figure 4.20 at least once, starting and finishing at node 1.

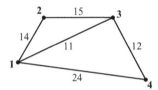

Figure 4.20

(ii) Find a shortest route that covers all of the arcs in Figure 4.20 at least once, starting and finishing at any suitable nodes.

[MEI adapted]

② Find a shortest route that covers all of the arcs in Figure 4.21 at least once, starting at Oxford and returning there. Distances are in miles.

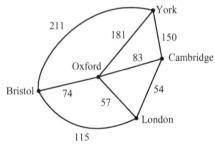

Figure 4.21

③ Find a shortest route that covers all of the arcs in Figure 4.21 at least once, starting at Oxford and ending at a different city.

④ Find a shortest route that covers all of the arcs in Figure 4.21 at least once, starting at one city and ending at a different one.

⑤ Find a shortest route that covers all of the arcs in Figure 4.22 at least once, starting at Okehampton and returning there.

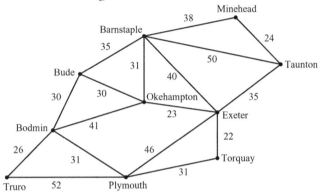

Figure 4.22

⑥ Find a shortest route that covers all of the arcs in Figure 4.22 at least once (you may start and finish at any town or city).

⑦ A highways maintenance depot must inspect all the manhole covers within its area. The road network is given below. In order to make the inspection an engineer must leave the depot, D, drive along each of the roads in the network at least once and return to the depot.

(i) What is the minimum distance that she must drive?

(ii) What route enables her to drive the distance in (i)?

(iii) How many times is node F visited during this route?

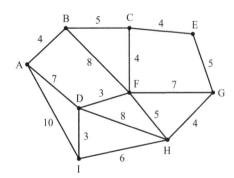

Figure 4.23

5 The travelling salesperson problem

In Chapter 2 a Hamiltonian cycle was defined as one that visits all of the nodes of a network exactly once and which returns to the starting point (another name for a Hamiltonian cycle is a **tour**). In the travelling salesperson problem, the aim is to find the shortest route that visits all of the nodes **at least once**. It may be that this route is a Hamiltonian cycle (i.e. where the nodes are not repeated), but the over-riding priority is for the length to be minimised.

Consider, for example, the networks in Figure 4.24 and Figure 4.25.

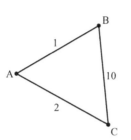

Figure 4.24   Figure 4.25

In Figure 4.24, a Hamiltonian cycle exists (e.g. ABCA), but you are better off with ABACA, in order to minimise the length.

In Figure 4.25, if A is to be the start and end point, then you cannot avoid repeating B, so that the route will not be a Hamiltonian cycle.

A problem is said to be **classical** if the aim is to find a Hamiltonian cycle (i.e. no node is to be repeated), and it is called **practical** if each node is to be visited **at least** once before returning to the start.

However, a practical problem can always be converted to a classical one by the following device.

- For each pair of nodes in the network, establish the shortest distance between them (which may be along an indirect path).
- Then create a complete graph with the given nodes and attach these shortest distances to the appropriate arcs.

For example, the network in Figure 4.24 becomes the one in Figure 4.26, whilst the network in Figure 4.25 becomes the one in Figure 4.27.

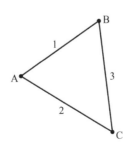

Figure 4.26   Figure 4.27

Note

The reason for the significance of the Hamiltonian cycle is that some of the algorithms that can be employed (and which you will be looking at shortly) are designed for Hamiltonian cycles.

Note

Whilst an upper bound corresponds to an actual solution that has been found, the lower bound is only a theoretical value below which the length cannot fall, and which might not be attainable in practice.

Unfortunately, there is no one algorithm that will enable you to find a shortest route. However, various methods exist for improving on a route that has already been found (i.e. one that passes through all the nodes, but is not of the shortest possible length).

The length associated with a route that has been found is called an **upper bound**, and so the aim is to reduce the upper bound.

There is also a method for finding a **lower bound** for the shortest distance. Once the upper bound is sufficiently close to the lower bound, you may decide that further effort is not worthwhile.

Finding an initial upper bound

Any Hamiltonian cycle provides an upper bound for the optimal solution to the travelling salesperson problem.

Example 4.7

The network showing the distances in miles between various cities is repeated in Figure 4.28. Find an upper bound for the travelling salesperson problem by comparing Hamiltonian cycles.

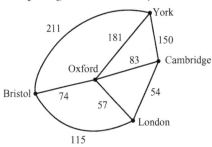

Figure 4.28

Solution

Three possible routes (with their total lengths) are as follows.

BYOCLB (211 + 181 + 83 + 54 + 115 = 644)
BYCOLB (211 + 150 + 83 + 57 + 115 = 616)
BOYCLB (74 + 181 + 150 + 54 + 115 = 574)

The upper bound (so far) is 574.

> The shortest length cannot be greater than 574, as you have already found a route of this length.

Finding a lower bound

You will now look at a method for finding a lower bound (often referred to as the **lower bound algorithm**). This method is only guaranteed to work if the network is complete.

Example 4.8

Use the lower bound algorithm to find a lower bound for a Hamiltonian cycle in the network shown in Figure 4.28.

Note

The lower bound is the combination of a minimum spanning tree and two arcs that join it to the excluded node. It is highly unlikely to be a tour, but, if it is, it will be the optimal solution.

Solution

Any lower bound for a Hamiltonian cycle will consist of two arcs from (say) B, together with three arcs linking O, L, C and Y.

The shortest possible total length of the two arcs from B is $74 + 115 = 189$.

To find the shortest possible total length of the arcs linking O, L, C and Y, you can find the minimum connector for these nodes.

By removing arcs in decreasing order, the length of the minimum connector for O, L, C and Y is found to be

$(181 + 150 + 83 + 57 + 54) - 181 - 83 = 261$.

Therefore the shortest possible total length of all the arcs (and a lower bound for a Hamiltonian cycle) is $189 + 261 = 450$.

So far, for the network shown in Figure 4.28, it has been established that the shortest possible length of a route that visits all the vertices lies between 450 and 574.

Further lower bounds can be established by dividing up the nodes differently. If node O is isolated instead of B, then the shortest possible total length of the two arcs from O is $57 + 74 = 131$, and using Kruskal's algorithm, for example, for the remaining nodes, you obtain a length of $54 + 115 + 150 = 319$, so that the lower bound is $131 + 319 = 450$ (again).

Isolating the other nodes in turn gives:

L + BOCY: $(54 + 57) + (74 + 83 + 150) = 418$

C + BOLY: $(54 + 83) + (57 + 74 + 181) = 449$

Y + BOLC: $(150 + 181) + (54 + 57 + 74) = 516$

The value of 516 supersedes the other, lower values: although it is true that the shortest route cannot be lower than 418, it is also true that it cannot be lower than 516.

You now have a lower bound of 516 and an upper bound of 574.

Using the nearest neighbour method to find an upper bound

This means that the method gives a possible tour.

The **nearest neighbour method** is a systematic way of finding a solution (and hence an upper bound). The method is only guaranteed to work if the network is complete.

Refer to Figure 4.28 again.

(i)   Start at any node (e.g. B).
(ii) Add the shortest arc leading to a new node: BO.
(iii) Repeat the process, to give BO + OL + LC + CY.

(iv) Return directly to the start node, to give the cycle:
BOLCYB (74 + 57 + 54 + 150 + 211 = 546)

(v) Repeat the algorithm, with other starting points.
OLCYBO (57 + 54 + 150 + 211 + 74 = 546)

LCOBY: can't return to L.

CLOBY: can't return to C.

YCLOBY (150 + 54 + 57 + 74 + 211 = 546)

In general, you take the lowest of the values obtained.

You now have a lower bound of 516 and an upper bound of 546.

Therefore the solution is written as

| This is not a solution if it does not give a tour. | → $516 \leqslant$ optimal solution $\leqslant 546$ ← | This will be a tour and it may be the optimal solution. |

It is easy to confuse Prim's algorithm with the nearest neighbour method. Make sure you know the difference.

Discussion point

→ What are the differences between the nearest neighbour method and Prim's algorithm?

The nearest neighbour method is a 'greedy' method: it doesn't look ahead, and just maximises the short-term gain, by selecting the nearest node. It doesn't usually give the best possible solution.

Solutions obtained by applying an algorithm can sometimes be improved.

A possible **tour improvement algorithm** can be illustrated by referring to Figure 4.28 again. The three possibilities that were mentioned earlier were:

BYOCLB (211 + 181 + 83 + 54 + 115 = 644)

BYCOLB (211 + 150 + 83 + 57 + 115 = 616)

BOYCLB (74 + 181 + 150 + 54 + 115 = 574)

Note that BYOCLB is improved by swapping O and C, or by swapping Y and O. The algorithm consists of examining each sequence of 4 nodes, and seeing if an improvement can be obtained by swapping the middle two nodes.

Using computer language, the algorithm could be written as follows (where N_i denotes the ith node, and d denotes distance between nodes).

For $i = 1$ to n
If $d(N_i, N_{i+2}) + d(N_{i+1}, N_{i+3}) < d(N_i, N_{i+1}) + d(N_{i+2}, N_{i+3})$ then swap N_{i+1} and N_{i+2}
Next i

Example 4.9

Apply the lower bound algorithm to the network in Figure 4.29, by isolating Exeter.

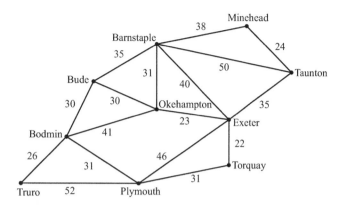

Figure 4.29

Solution

Exclude Exeter and all arcs connected to it.

Then find a minimum connector for the nodes that remain.

You can remove arcs in decreasing order, as follows (ensuring that the network remains connected).

Tru–Ply 52

Bar–Tau 50

Bod–Oke 41

Bud–Bar 35

This leaves the following arcs in the minimum connector.

Tor–Ply 31

Ply–Bod 31

Bod–Tru 26

Bod–Bud 30

Bud–Oke 30

Oke–Bar 31

Bar–Min 38

Min–Tau 24

The total weight is 241.

The two shortest arcs leading from Exeter are 22 and 23: add these to the total weight of the minimum connector.

So the lower bound for the length of the tour is 241 + 22 + 23 = 286.

Note

The lower bound is made as high as possible (i.e. as close to the upper bound as possible) to reduce the interval in which the optimal solution lies.

If the lower bound is equal to the upper bound, then the solution is optimal.

ACTIVITY 4.9

Apply the nearest neighbour method to the network in Figure 4.29, with Taunton as the starting point.

① Apply the lower bound algorithm to the network in Figure 4.18 (repeated in Figure 4.30), isolating each of the nodes in turn. What is the lower bound that it produces?

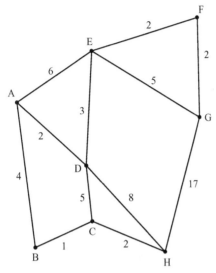

Figure 4.30

② Apply the nearest neighbour method to the network in Figure 4.30, considering each of the nodes as a possible starting point. What is the upper bound that it produces?

③ Apply the lower bound algorithm to the network in Figure 4.31 (from question 3 of Exercise 4.1), isolating each of the nodes in turn. What is the lower bound?

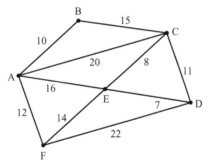

Figure 4.31

④ Apply the nearest neighbour method to the network in Figure 4.31, considering each of the nodes as a possible starting point.

⑤ For the network in Figure 4.32, create the table of weights associated with the complete network obtained by finding the shortest distance between each pair of nodes (i.e. converting a practical problem to a classical problem).

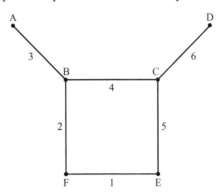

Figure 4.32

⑥ A group of tourists staying in Weston wishes to visit all the places shown on the following map.

(i) Suggest a route that will minimise their total driving distance.

(ii) A tree has blocked the road between Weston and Cheddar, making it impassable for the whole day. Suggest an alternative route that will result in the least extra driving distance.

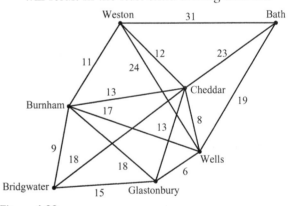

Figure 4.33

⑦ A depot located in Birmingham supplies goods to customers in Sheffield, Nottingham, Stoke, Shrewsbury, Hereford, Gloucester and Northampton.

(i) Plan a suitable route for the delivery lorry if it has to make deliveries in all of these towns on one trip. The distances involved are shown in the table.

(ii) Suggest two reasons why the shortest route may not be the best route for the lorry.

	Birmingham	Sheffield	Nottingham	Stoke	Shrewsbury	Hereford	Gloucester	Northampton
Birmingham	–	77	50	43	43	52	52	50
Sheffield	77	–	37	47	79	125	128	94
Nottingham	50	37	–	50	79	102	102	57
Stoke	43	47	50	–	34	83	89	85
Shrewsbury	43	79	79	34	–	52	75	93
Hereford	52	125	102	83	52	–	28	91
Gloucester	52	128	102	89	75	28	–	72
Northampton	50	94	57	85	93	91	72	–

Table 4.9

Solving network problems involves selecting an appropriate algorithm.
The algorithm may need adapting to meet the requirements of the problem. It may
also need adapting to take into account practical constraints.

Exercise 4.5

① A network has four odd vertices, A, B, C and D.
One pairing of them is A with B and C with D.

(i) How many ways are there of pairing four odd vertices?

(ii) How many ways are there of pairing six odd vertices?

(iii) How many ways are there of pairing eight odd vertices?

(iv) How many ways are there of pairing twenty odd vertices?

② The graph below is a representation of a system of roads. The lengths of the roads are shown in metres.

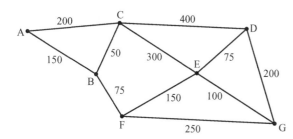

Figure 4.34

(i) List the odd vertices in the graph.

(ii) Explain why the graph is not Eulerian.

(iii) Find the shortest route that starts and finishes at A and traverses each road at least once. State the length of the route.

③ In a sweet-making factory, five flavours of fruit drop are made one after another on a single machine. After each flavour, the machine must be cleaned in readiness for the next flavour. The time spent cleaning depends on the two flavours as indicated in the table below.

Time in minutes	Next flavour to be made				
Last flavour made	Strawberry	Lemon	Orange	Lime	Raspberry
Strawberry	–	14	12	19	16
Lemon	21	–	14	10	19
Orange	19	16	–	17	20
Lime	17	9	13	–	15
Raspberry	20	15	13	19	–

Table 4.10

The production manager wishes to find a sequence that minimises the total time spent cleaning the machine in each cycle from strawberry to strawberry, making each flavour of fruit drop once only per cycle.

(i) By constructing an appropriate network, explain how the problem may be formulated as a travelling salesperson problem. Hence, by using the nearest neighbour algorithm starting from strawberry, suggest a production sequence to the manager.

Someone notices that the smallest cleaning time is in changing from lime to lemon. Accordingly, he suggests that a better production sequence may be found by using the nearest neighbour algorithm starting from lime, so that the sequence will begin with the change from lime to lemon.

(ii) Determine whether he is right that a better sequence will be found.

④ A depot located at town A supplies goods to customers in towns B, C, D and E.

The inter-town distances are given in the table below.

	A	**B**	**C**	**D**	**E**
A	–	28	57	20	45
B	28	–	47	46	73
C	57	47	–	76	85
D	20	46	76	–	40
E	45	73	85	40	–

Table 4.11

Usually a single vehicle will suffice for a particular delivery but today the customers' requirements are 100 units each and the vehicle available will only carry 300 units. Another similar vehicle can be hired locally. Explain how the two vehicles should be routed.

(Hint: introduce an artificial depot.)

⑤ (i) When moored alongside a harbour wall a sailing boat has to be secured by four ropes. The ropes are attached to the points A, B, C and D shown in the diagram.

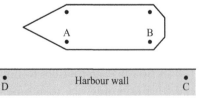

Figure 4.35

The point A has to be attached to D and also to C. The point B has to be attached to C and also to D.

(a) Regard the mooring ropes as defining the arcs of a network and the points A, B, C and D as being the nodes. Give the orders of the nodes.

(b) Explain, in terms of the orders of the nodes, why the network is Eulerian. Indicate the implications of this if only one long length of rope is available instead of four shorter lengths.

(ii) When mooring alongside another boat, ropes are attached both to the other boat and also to the harbour wall, as shown.

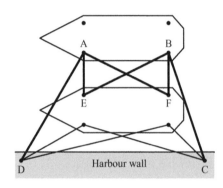

Figure 4.36

Regard the ropes marked with bold lines in the diagram, together with the points A, B, C, D, E and F, as a network.

(a) Give the orders of the nodes and explain why the existence of odd nodes means that the network is not Eulerian.

(b) What practical consequence does the non-Eulerian nature of the network have on the use of a single length of rope to moor the boat?

(c) The solution to the route inspection problem for the network gives a way of mooring the boat with a single length of rope.

Solve the route inspection problem for the network, starting and finishing at the same node, and give the corresponding sequence of nodes.

(d) Suppose that in part (c), the route is allowed to start at C and finish elsewhere. Give two solutions, both starting at C but finishing at different nodes, in which no more than two arcs are repeated.

⑥ The following matrix gives the costs of tickets for direct flights between six connected cities.

From \ To	A	B	C	D	E	F
A	–	45	60	58	90	145
B	45	–	67	25	83	100
C	60	70	–	50	70	320
D	50	25	50	–	35	210
E	100	80	70	35	–	72
F	145	110	300	175	80	–

Table 4.12

In this question a tour is a journey from a city, visiting each other city once and only once, and returning to the starting city.

(i) Use the nearest neighbour method to find a tour, starting and finishing at A, with a low associated cost.

Show that the method has not produced the minimum cost tour.

(ii) How many different tours that start and finish at A are there altogether?

(iii) Suppose that, in addition to the cost of tickets, airport taxes must be paid on leaving an airport according to the following table. Costs are in £.

A	B	C	D	E	F
20	30	20	40	10	20

Table 4.13

Thus the flight from A to B will cost £45 for the ticket and £20 tax, giving a total of £65.

Produce a matrix showing total costs, i.e. the total of fares and taxes.

(iv) Give an example where the cheapest route from one city to another differs when taxes are taken into account from that when taxes are not taken into account. Show your two routes and give their costs.

(v) Do airport taxes have an effect on the problem of finding the cheapest tour starting and finishing at A? Why?

[MEI]

⑦ White, yellow, blue, green and red dyes are to be used separately in a dyeing vat (a container in which materials are dyed). Each colour is to be used once during each day. The vat has to be cleaned between colours, and the cost of this depends on which colour was previously used, and on which colour is going to be used next. For example, if the blue dye was previously used, and the yellow dye is going to be used next, the cost is 5. These costs, in suitable units, are shown in the table. The vat must be returned to its original colour at the end of the day.

From \ To	W	Y	B	G	R
W	–	0	2	1	2
Y	4	–	4	3	4
B	8	5	–	1	2
G	7	3	1	–	3
R	7	4	3	3	–

Table 4.14

(i) Explain why this problem is similar to the travelling salesperson problem.

(ii) Use the nearest neighbour method five times, starting from each colour in turn, to find a low-cost sequence of colours.

(iii) Give a colour sequence of cost 12, starting and ending with white.

(iv) Because the network is directed, the technique of deleting a node and finding a minimum connector for the remainder to produce a lower bound will not work. Explain why not.

[MEI]

LEARNING OUTCOMES

Now you have finished this chapter, you should be able to

➤ understand and use the language of networks, including: node, arc and weight

➤ solve network optimisation problems using spanning trees

➤ find the shortest route between two nodes of a network using Dijkstra's algorithm

➤ solve route inspection problems

➤ find and interpret upper bounds and lower bounds for the travelling salesperson problem

➤ evaluate, modify and refine models that use networks.

KEY POINTS

1 A network is a weighted graph.

2 The minimum connector problem is solved by creating a spanning tree with the minimum weight. The following methods can be employed.
   - Remove arcs in order of decreasing weight.
   - Apply Prim's algorithm.
   - Apply Kruskal's algorithm.

3 Prim's algorithm can be applied to the table of weights.

4 Prim's and Kruskal's algorithms have cubic complexity.

5 Dijkstra's algorithm provides a procedure for determining the shortest route between two particular nodes of a network.

6 Dijkstra's algorithm has quadratic complexity.

7 The route inspection problem is to find the shortest route that covers all of the arcs at least once – returning to the starting point, if required. The problem is solved by classifying networks as:
   - Eulerian if they have no odd nodes
   - semi-Eulerian if they have two odd nodes
   - neither Eulerian nor semi-Eulerian if they have four or more odd nodes.

8 A formula can be established for the number of possible pairings for the route inspection problem.

9 The travelling salesperson problem is to find the shortest route that visits all of the nodes of a network at least once, returning to the starting point.

10 A problem is said to be classical if you aim to find a Hamiltonian cycle, and is practical otherwise. A practical problem can always be converted into a classical one.

11 The shortest possible route can be placed within bounds by applying the lower bound algorithm and the nearest neighbour algorithm.

12 Computers may employ a tour improvement algorithm.

5

Critical path analysis

Let all things be done decently and in order.

The First Epistle of Paul to the Corinthians, xiv 40

→ Each morning Paul makes toast using a grill. The grill can take two slices of bread at a time and takes 1 minute to toast each side of the bread. How long does it take Paul to toast three slices of bread?

1 Constructing a network

Critical path analysis uses networks to help schedule projects involving a number of activities, some of which require other activities to take place before they can begin.

The starting point is the **precedence table** (also known as a dependence table), as shown in Table 5.1.

Activity	Immediately preceding activities	Duration (in hours)
A	–	4
B	–	5
C	A, B	6
D	B	2
E	B	4
F	C, D	3
G	C, D, E	1

Table 5.1

From this you can construct a network that places the activities in the correct relation to each other; i.e. taking account of the precedences.

There are two conventions that can be adopted for the network. The one that you will employ in this chapter is called **activity–on–arc**. The alternative convention is called activity–on–node.

Example 5.1

Draw the network for the precedence table shown in Table 5.1.

Solution

The network is shown in Figure 5.1.

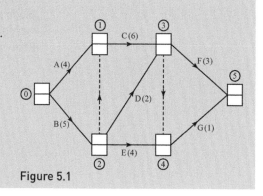

Figure 5.1

The activities (on the arcs) lead into the nodes, which are referred to as **events**, and are assigned two boxes, which will be explained shortly. The dotted lines, referred to as **dummy activities**, will also be explained shortly.

Events can only start once the activities that lead into them have been completed. For example, activity F can only start once activities C and D have been completed, and the event labelled 3 takes account of this.

An event is described as a **merge event** if it has two or more incoming activities (e.g. event 3 in Figure 5.1), and as a **burst event** if it has two or more outgoing activities (e.g. event 2).

Activity B needs to be completed before C, D and E can start, but this creates a problem. Suppose that the arcs are connected as shown in Figure 5.2.

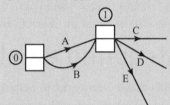

Figure 5.2

Although B leads into C, D and E, as required, Figure 5.2 implies that D and E depend on A, which is not correct.

Another issue is that, in Figure 5.2, A and B share both a start node and an end node. This creates a problem because activities are usually identified by their start and end nodes (by a computer, for example).

Both of these problems are solved by creating a dummy arc leading from event 2 to event 1. Informally, you can think of a message being sent from event 2 to event 1, to say that the activities leading into event 2 have been completed.

The dummy arc leading from event 3 to event 4 represents the fact that G depends on both C and D, as well as E.

Dummy activities have duration zero.

Note that event 5 has been created to represent the end of the project. Only activities F and G need lead into it, as the other activities all lead into F or G, either directly or indirectly.

ACTIVITY 5.1

How would the network be changed if activity G is no longer dependent on activity C?

Note

Whereas the route inspection and travelling salesperson problems involve an individual (for example) travelling through the network, so that they can only be in one place at a time, critical path analysis typically involves a number of people working simultaneously on different activities, so that different parts of the network are in operation at the same time.

Determining earliest and latest event times

You are usually concerned with how soon the whole process can be completed. This is determined by making a **forward pass** through the network, recording the **earliest** (or **early**) **event times**.

Look at the network in Figure 5.1.
- Event 0 is given an earliest event time of 0.
- Event 1 cannot start until 5 hours have elapsed, in order that both the activities A and B have been completed. So 5 is entered in the upper box for event 1.
- 5 is entered in the upper box for event 2, as this only depends on activity B.
- Event 3 has to wait until 6 hours after event 1 has started, and also 2 hours after event 2 has started: i.e. not before $5 + 6 = 11$ and $5 + 2 = 7$ hours have elapsed; i.e. 11 hours.
- Event 4 has to wait until 4 hours after event 2 has started, and also until event 3 has started: i.e. not before $5 + 4 = 9$ and 11 hours have elapsed; i.e. 11 hours.
- Finally, event 5 has to wait until 3 hours after event 3 has started, and also until 1 hour after event 4 has started: i.e. not before $11 + 3 = 14$ and $11 + 1 = 12$ hours have elapsed; i.e. 14 hours.

Figure 5.3 shows the situation once all the earliest event times have been determined. The whole process can be completed in 14 hours. This is the **minimum completion time** (or **critical time**).

> ## Note
> The early event time for a merge event cannot be determined until the early event times for all the events for the incoming arcs have been assigned. For example, event 2 needs its early event time before 5 can be determined as the earliest time for event 1.

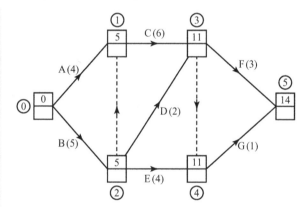

Figure 5.3

Some of the activities may have some slack available. For example, A could be delayed by up to 1 hour without increasing the duration of the project. Those activities that do not have any slack available are called **critical activities**.

In order to establish the extent of any available slack, you carry out a **backward pass**, starting at the end node.

- Place a 14 in the lower box for event 5, to indicate its **latest** (or **late**) **event time**.
- As activity G has a duration of 1, the latest time at which event 4 could end, in order to complete event 5 by time 14, is 13, which is therefore the latest event time.
- For event 3 it is 11, in order to complete event 4 by time 13 and event 5 by 14.

- For event 1 it is 5, in order to complete event 3 by time 11.
- For event 2 it is 5, in order to complete event 1 by time 5.
- Finally, the latest event time for event 0 is 0.

Figure 5.4 shows the situation once all the latest event times have been determined.

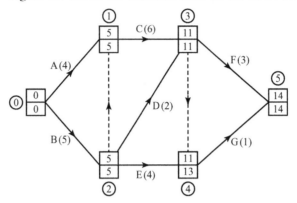

Figure 5.4

Having found the earliest and latest event times (EET and LET), you can now establish the **earliest start and latest finish times** for the activities (be careful not to confuse these similar sounding terms). Note that the latest finish time means the latest time by which the activity must finish, in order not to affect the project completion time.

First, you can identify each activity by its start node i and end node j: writing (i, j).

Consider the part of a network involving activity (i, j), shown in Figure 5.5.

Figure 5.5

The earliest start time for (i, j) is 4 (in general, EET_i).

The latest finish time for (i, j) is 13 (in general, LET_j).

The (total) **float** of an activity is the available slack, and is equal to

(latest finish time – earliest start time) – duration of activity.

In the example above, the float is $(13 - 4) - 2 = 7$.

① Draw an activity network to represent this project.

Activity	Immediate predecessors	Duration
A	–	5
B	A	2
C	–	8
D	A, C	12
E	B, C	6

Table 5.2

② The table shows the precedences for the four tasks of a project. The duration of each task is also shown.

Task	Immediate predecessors	Duration (days)
A	–	2
B	–	1
C	A	1
D	A, B	3

Table 5.3

Draw an activity-on-arc network for the project, showing the earliest and latest event times. Give the minimum completion time.

③ Draw the activity-on-arc network for the project with the activities listed below.

Show the earliest and latest event times.

Activity	Immediate predecessors	Duration (days)
A	–	8
B	–	4
C	A	2
D	A	10
E	B	5
F	C, E	3

Table 5.4

④ The table shows the activities involved in building a short length of road to bypass a village. The table gives their durations and their immediate predecessors.

	Activity	Immediate predecessors	Duration (weeks)
A	Survey sites	–	8
B	Purchase land	A	22
C	Supply materials	–	10
D	Supply machinery	–	4
E	Excavate cuttings	B, D	9
F	Build bridges and embankments	B, C, D	11
G	Lay drains	E, F	9
H	Lay hardcore	G	5
I	Lay bitumen	H	3
J	Install road furniture	E, F	10

Table 5.5

Draw an activity-on-arc network for these activities, showing the earliest and latest event times. Give the minimum completion time.

⑤ A construction project involves nine activities. Their immediate predecessors and durations are listed in the table.

Activity	Immediate predecessors	Duration (days)
A	–	5
B	–	3
C	–	6
D	A	2
E	A, B	3
F	C, D	5
G	C	1
H	E	2
I	E, G	4

Table 5.6

Draw an activity-on-arc network for the project, showing the earliest and latest event times. Give the minimum completion time.

2 Solving problems

When a project does not go according to plan or new constraints are added, the network needs to be revised and the schedule for the activities changes. The float is the first thing to consider when this happens.

The float can be broken down into two parts.

Referring to Figure 5.5:

The **independent float**: $(10 - 6) - 2 = 2$.

In general, independent float = $(EET_j - LET_i)$ − duration of activity (or zero, if this is negative).

This is the component of the float that is guaranteed, in the sense that it is not affected by other activities starting late or over running (provided that the project is still completed on time).

The remainder of the total float is described as the **interfering float**: i.e. $7 - 2 = 5$, in this case.

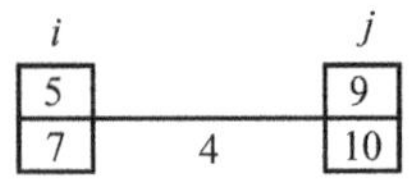

Figure 5.6

Figure 5.6 shows an example where the independent float is initially calculated as $(9 - 7) - 4 = -2$, and then set to zero.

The total float is $(10 - 5) - 4 = 1$, and this is all allocated to the interfering float.

In this case, there is some slack for the activity, but it is dependent on the extent to which other activities start late or overrun. The initially negative independent float means that the project may overrun if other activities take advantage of their own floats. For example, if the activity doesn't start before time 7, then it won't be completed by the latest finish time of 10.

A **critical activity** has zero float.

This will only be possible if
$LET_i = EET_i$, $LET_j = EET_j$, and $EET_j = EET_i$ + duration.

Example 5.2

Establish the critical activities and floats for the network shown in Figure 5.4, breaking the floats down into independent and interfering floats.

Solution

A: float of $5 - 0 - 4 = 1$ (independent: 1, interfering: 0)

B: float of $5 - 0 - 5 = 0$; critical activity

C: float of $11 - 5 - 6 = 0$; critical activity

D: float of $11 - 5 - 2 = 4$ (independent: 4, interfering: 0)

E: float of $13 - 5 - 4 = 4$; (independent: 2, interfering: 2)

F: float of $14 - 11 - 3 = 0$; critical activity

G: float of $14 - 11 - 1 = 2$ (independent: 0, interfering: 2)

The critical activities B, C and F form a path through the network, leading from the start to the end. Such a **critical path** will always exist, though there may be more than one.

The length of the critical path is the minimum completion time of the project. If there is more than one such path, they will all have the same length. These paths are the longest possible paths in the network (Linear programming methods can use this property to solve critical path analysis problems).

ACTIVITY 5.2

Given that all of the activities in the network shown in Figure 5.4 overrun by 1 hour, find the new minimum completion time for the project.

It may be that the completion time of the project has to be reduced, and that this can be achieved by using extra workers to reduce the durations of the critical activities. This is sometimes referred to as **crashing the network**. Any such reductions may change the activities that make up the critical path.

Example 5.3

Table 5.7 shows the costs of reducing the durations of each of the activities in Example 5.1 by one hour, as well as the minimum duration that is possible for each activity.

Activity	Original duration (in hours)	Cost of reducing duration by 1 hour	Minimum duration possible
A	4	£100	2
B	5	£200	4
C	6	£100	3
D	2	£300	1
E	4	£200	1
F	3	£200	2
G	1	–	1

Table 5.7

Suppose that it is necessary to complete the project as quickly as possible, and you wish to know the extra cost involved in doing this.

Solution

One way of tackling this problem is to repeat the original process, using the minimum durations, and then to increase the durations of any non-critical activities.

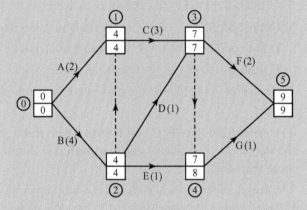

Figure 5.7

The new minimum completion time is 9 hours.

A: float of 2

B: critical activity

C: critical activity

D: float of 2

E: float of 3

F: critical activity

G: float of 1

You can now increase the durations of the non-critical activities, so as not to incur unnecessary costs. The results are shown in Table 5.8. Note that, for activities D and E, there is no need to increase the duration above the original figures.

Activity	Original duration (in hours)	Cost of reducing duration by 1 hour	Minimum duration possible	New duration	Extra cost
A	4	£100	2	4	£0
B	5	£200	4	4	£200
C	6	£100	3	3	£300
D	2	£300	1	2	£0
E	4	£200	1	4	£0
F	3	£200	2	2	£200
G	1	–	1	1	£0

Table 5.8

The total extra cost is £700.

① A project is described in the table below.

Activity	Immediate predecessors	Duration (days)
A	–	8
B	–	4
C	A	2
D	A	10
E	B	5
F	C, E	3

Table 5.9

The duration of any of the activities can be reduced at a cost, as given in the table below.

Activity	Cost (£) for each day by which duration is reduced	Minimum duration
A	50	2
B	100	2
C	40	1
D	60	5
E	25	4
F	10	2

Table 5.10

Which activity durations should be reduced, and by how many days, to reduce the minimum project completion time by

(i) 2 days   (ii) 7 days

at minimum cost?

② Find the total float for each activity in this project.

Activity	Immediate predecessors	Duration
A	–	2
B	–	1
C	A	1
D	A, B	3

Table 5.11

[MEI adapted]

③ The activities involved in cooking a meal of toad-in-the-hole, potatoes and cabbage, followed by apple pie and custard, are shown in the table below.

Activity		Immediate predecessors	Duration (mins)
A	Grill sausages	–	8
B	Make batter	–	6
C	Make apple pie	–	15
D	Prepare potatoes	–	6
E	Prepare cabbage	–	4
F	Cook sausages and batter together	A, B	35
G	Cook potatoes	D	25
H	Cook cabbage	E	8
I	Cook apple pie	C	30
J	Make custard	–	8

Table 5.12

(i) Draw an activity network to represent this project.

(ii) Find the minimum project completion time, assuming that there are enough people available to carry out the activities.

(iii) Which activities are critical?

3 Cascade charts

In addition to establishing the critical activities of an operation, it will usually be important to take account of the resources (people and equipment) required at different stages, and to try to make use of them in the most efficient way.

A cascade chart shows how the activities in a project can be scheduled in relation to each other. It can be drawn from the information in the activity network and shows the critical activities and the float on the non–critical activities.

Example 5.4

Draw a **cascade** (or Gantt) **chart** to display the activities shown in the network in Figure 5.4 on page 92.

Solution

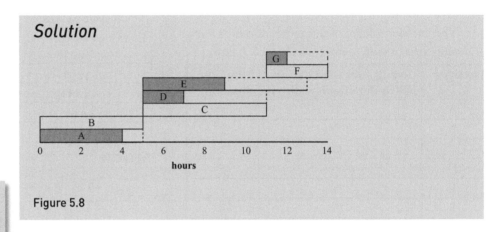

Figure 5.8

Note

Not all of the information about precedences can easily be reflected in the cascade chart. For example, activity E has to precede activity G, so if E is delayed until the last moment, it will also be necessary to delay G until the last moment. Vertical dotted lines are sometimes introduced, to show the constraints on activities, but they may be difficult to interpret.

Cascade charts can be constructed in slightly different ways, but generally have the following features.

- Activities are represented by horizontal bars, beginning at their earliest start times.

- Each activity should be drawn on a separate line. However, critical activities can be drawn on a single horizontal line.

- The bars are divided into two parts: the length of the first part is the duration of the activity, whilst the length of the second part is its float (the second part can be indicated by shading or a dotted border).

- The number of workers required for an activity may be shown on the bar.

One use of the cascade chart is to establish which activities definitely have to be taking place at a given point in time: i.e. activities that can't be shifted away from a particular vertical line.

A **resource histogram** can be drawn, based on the cascade chart. It shows the number of workers required at any given time.

Example 5.5	Suppose that the activities of the network shown in Figure 5.4 on page 92 require the following numbers of workers.

A: 2   B: 3   C: 1   D: 5   E: 2   F: 4   G: 2

Draw the resource histogram based on the cascade chart shown in Figure 5.8.

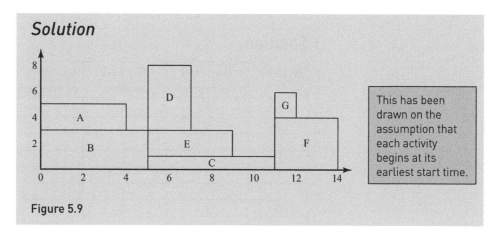

Solution

This has been drawn on the assumption that each activity begins at its earliest start time.

Figure 5.9

A scheduling diagram can also be constructed. This shows which activity each worker should be performing throughout the project.

Resource problems

A factory is likely to employ a workforce on a full-time basis and will want to keep its workers occupied, as far as possible. It wants to avoid taking on expensive temporary workers. For these reasons, it is desirable to smooth out the number of workers required over the duration of a project. This is referred to as **resource levelling**. It is an example of a **heuristic procedure** (when the solution is likely to be reasonably good, but is not guaranteed to be optimal).

The first step is to examine the resource histogram, to see if the floats allow activities to be shuffled around, in order to limit the number of workers required without exceeding the critical time.

In the case of Figure 5.9 (relating to Example 5.5), you can see that there is no scope for reducing this limit below six, without extending the duration of the project. In other words, a workforce of six is needed, if temporary workers are not to be employed.

The lowest possible value for this limit can be established by considering areas in the resource histogram. The best you can hope for is that the resource levelling produces a rectangle, with an area equal to the sum of the areas of the blocks in the original resource histogram. This sum is

$$\Sigma \text{ (duration of activity} \times \text{number of workers required for the activity)}$$

where Σ denotes summation over all the activities.

The height of the (ideal) rectangle is then obtained by dividing the area by the base, which is the critical time.

Example 5.6

Find the lowest possible value for the workforce needed in the case of Table 5.13.

Solution

Activity	Duration (in hours)	Workers needed	Duration × number of workers
A	4	2	8
B	5	3	15
C	6	1	6
D	2	5	10
E	4	2	8
F	3	4	12
G	1	2	2
			Total = 61

Table 5.13

This is the same project that was looked at in Example 5.1. The minimum completion time is 14 hours.

The area of the ideal rectangle is $8 + 15 + 6 + 10 + 8 + 12 + 2 = 61$.

Dividing by the critical time gives $61 \div 14 = 4.36$

so a workforce of 5 is the smallest possible number needed.

> However, as you have seen, the actual workforce needed is 6. This is close to the lower bound and may be acceptable so as not to increase the duration.

Exercise 5.3

① Harry wants to decorate a room. The activities involved are given in the table, together with their durations and immediate predecessors.

Activity		Immediate predecessors	Duration (days)
A	Remove old wallpaper	–	1
B	Prepare wooden surfaces	–	0.25
C	Paint ceiling	A	0.75
D	Apply undercoat	A, B	1
E	Apply gloss paint	D	1
F	Paper walls	C, E	1

Table 5.14

(i) Represent the project as an activity network.

Harry decides to decorate three rooms. The activities, durations and predecessors for each room are as given in the table. Harry will be helped by his friend Nisha, but activities cannot be shared between Harry and Nisha.

(ii) Show how Harry and Nisha can decorate the three rooms when they only have 8 days available.   [MEI adapted]

②

Activity	A	B	C	D	E	F	G	H	I	J	K	L	M	N
Duration (days)	1	2	4	3	14	14	16	12	14	10	5	4	6	3
Immediate predecessors	–	A	B	A	C	C	C	D G	D G	D G	H	I	J	E F K L M
Number of workers	1	1	1	1	1	2	2	2	3	1	1	2	1	2

Table 5.15

(i) Calculate the minimum time in which this project can be completed and determine which activities are critical.

(ii) Calculate the float of the non-critical activities, distinguishing between the independent and the interfering float.

(iii) Draw a cascade chart for the project.

(iv) Describe in detail the effect on the project if only five workers are available, each of whom can carry out any of the activities.

③ The table shows activities involved in a construction project, their durations, and their immediate predecessors.

	Activity	Immediate predecessors	Duration (weeks)
A	Obtain planning permission	–	6
B	Survey site	–	2
C	Dig foundations	A, B	6
D	Lay drains	A, B	3
E	Access work	A	10
F	Plumbing	C, D	2
G	Framework	C	6
H	Internal work	E, F, G	4
I	Brickwork	G	3

Table 5.16

(i) Draw an activity network for the project.

(ii) Perform a forward pass and a backward pass to find earliest and latest event times. Give the critical path and the minimum time to complete the project.

(iii) The contractor winning the contract has only one JCB (a digging machine) available. This is needed for activities C (digging foundations) and D (laying drains). Decide whether or not the contractor can complete the project within the minimum time found in (ii). Give reasons and working to support your conclusion. [MEI]

④ The table shows the activities involved in a project, their durations, and their immediate predecessors.

Activity	A	B	C	D	E	F	G	H
Duration (days)	1	2	1	1	4	2	3	2
Immediate predecessors	–	–	A	B	B	C, D	E, F	E

Table 5.17

(i) Draw an activity-on-arc network for the project.

(ii) Perform a forward pass and a backward pass on your network to determine the earliest and latest event times. State the minimum time for completion and the activities forming the critical path.

(iii) Draw a cascade chart for the project, given that all the activities are scheduled to start as early as possible.

(iv) The number of people needed for each activity is as follows.

Activity	A	B	C	D	E	F	G	H
People	1	4	2	3	1	2	3	2

Table 5.18

Activities C and F are to be scheduled to start later than their earliest start times so that only five people are needed at any one time, whilst the project is still completed in the minimum time. Specify the scheduled start times for activities C and F. [MEI]

LEARNING OUTCOMES

Now you have finished this chapter, you should be able to

➤ construct, represent and interpret a precedence (activity) network using activity-on-arc

➤ determine earliest and latest event times for an activity network

➤ determine earliest start and latest finish times for an activity

➤ identify critical activities, critical paths and the floats of non-critical activities

➤ refine models and understand the implications of possible changes in the context of critical path analysis

➤ construct and interpret cascade (Gantt) charts and resource histograms

➤ carry out resource levelling (using heuristic procedures) and evaluate problems where resources are restricted.

KEY POINTS

1 Critical path analysis uses networks to help schedule projects involving a number of activities.

2 Based on a precedence table, an activity-on-arc network is constructed.

3 Forward and backward passes are made to determine earliest and latest event times.

4 A critical path is established, in order to find the minimum completion time.

5 Extra resources may be employed to reduce the durations of the critical activities.

6 Cascade charts and resource histograms can be used to display the activities.

7 Resource levelling may be employed to limit the number of workers used.

6 Linear programming

Tout est pour le mieux dans le meilleur des mondes possibles.
Voltaire, Candide

→ Think about a product that you might make to sell at a fundraising event. What factors affect how much profit you can make?

1 Formulating and solving constrained optimisation problems

Formulating constrained optimisation problems

Linear programming is a way of solving logistical problems involving linear constraints, as in the following example.

Example 6.1

Millie bakes some cakes for a village fete. She makes cupcakes and chocolate cakes. She sells both types of cake for £1 each.

Each cupcake uses 10 g of sugar and 30 g of flour, and each chocolate cake uses 15 g of sugar and 25 g of flour.

She has only 60 g of sugar and 150 g of flour, and wants to make as much money as possible.

Formulate this as a linear programming problem.

> This is a **constrained optimisation** problem, and the first step is to formulate the constraints mathematically.

Solution

Let x be the number of cupcakes to be made, and y the number of chocolate cakes. The two constraints involving the ingredients become

$2x + 3y \leq 12$ (sugar) ← The constraint is simplified from $10x + 15y \leq 60$.

$6x + 5y \leq 30$ (flour). ← The constraint is simplified from $30x + 25y \leq 150$.

Also, $x \geq 0, y \geq 0$. ← The values of x and y can't be negative.

The total value of sales is $x \times 1 + y \times 1$, and so

$P = x + y$ is the **objective function** that you wish to maximise.

The variables x and y are referred to as the **control variables**.

Solving constrained optimisation problems

The cake problem from Example 6.1 can be represented graphically, as shown in Figure 6.1.

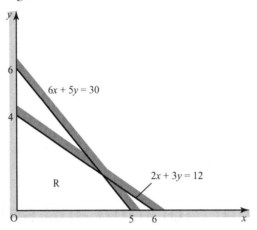

Figure 6.1

The region R in Figure 6.1 is the **feasible region**, where all the constraints are satisfied.

Note that it is the unwanted areas that are shaded. This makes it easier to identify the feasible region – especially where there are a number of constraints.

You now find which points of the feasible region maximise P.

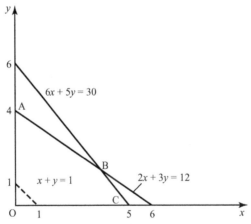

Figure 6.2

> **Note**
>
> It can help to use a ruler to represent the moving line, but if it still isn't obvious which vertex is required, you can evaluate the objective function at each vertex, to find the one with the largest value for P.

In Figure 6.2, the line $P = x + y$ is parallel to $x + y = 1$, and needs to be as far away from O as possible, in order to maximise P.

You need to visualise the objective line $x + y = 1$ being moved away from O, keeping its gradient the same, and find the point at which it is about to leave the feasible region. Therefore the maximum value of P will occur at one of the **vertices** (i.e. corners) of the feasible region (unless the line $P = x + y$ has the same gradient as one of the constraint lines). In this example it is vertex B.

Example 6.2

Find the optimal solution for the following linear programming problem.

Maximise $P = x + y$
subject to $2x + 3y \leqslant 12$
$\qquad\qquad 6x + 5y \leqslant 30$
$\qquad\qquad x \geqslant 0, y \geqslant 0$.

Solution

Look at Figure 6.2.

At B, the lines $6x + 5y = 30$ ① and $2x + 3y = 12$ ② intersect.

Solve these simultaneously.

$3 \times ② - ① \Rightarrow 4y = 6 \Rightarrow y = \frac{3}{2}$

Then $② \Rightarrow 2x = 12 - \frac{9}{2} \Rightarrow x = 6 - \frac{9}{4} = \frac{15}{4}$

$\Rightarrow P = x + y = \frac{21}{4} = 5.25$ at B $(3.75, 1.5)$.

At A, $P = 4$, and at C, $P = 5$, which confirms that B is the required vertex, so the optimal solution is $x = 3.75$ and $y = 1.5$.

In some contexts, non-integer solutions may be possible.

However, for the problem given in Example 6.1, only integer values are acceptable as x and y represent numbers of cakes. In this case, you can consider integer points neighbouring $(3.75, 1.5)$, provided they are within the feasible region. This is a method that is quick to apply. However, it may not give the optimal solution. A more powerful approach is the 'branch-and-bound' method, and this is considered in the next section.

Example 6.3

Solve the linear programming problem given in Example 6.2 if the values of x and y can only take integer values.

Solution

The constraints are that

$6x + 5y \leqslant 30$ and $2x + 3y \leqslant 12$.

Maximise $P = x + y$.

Considering integer points neighbouring $(3.75, 1.5)$ gives

$(3, 1)$: $6x + 5y = 23$ and $2x + 3y = 9$; $P = 4$

$(3, 2)$: $6x + 5y = 28$ and $2x + 3y = 12$; $P = 5$

$(4, 1)$: $6x + 5y = 29$ and $2x + 3y = 11$; $P = 5$

$(4, 2)$: $6x + 5y = 34$ (reject)

> $(4, 2)$ is not in the feasible region.

The points $(3, 2)$ and $(4, 1)$ give equally good solutions. However, that does not guarantee that this is the optimal solution.

> Always state the solution in terms of the context of the problem.

So, in order to maximise the total value of sales, the output of cakes should be either 3 cupcakes and 2 chocolate cakes, or 4 cupcakes and 1 chocolate cake.

> **Note**
> ------------
> $(5, 0)$, vertex C, also gives $P = 5$, so 5 cupcakes is another solution that maximises the objective function, but with only one type of cake!

Instead of maximising sales or profit, you may want to minimise costs, for example. It may be possible to create a feasible region in this case, depending on the constraints.

Example 6.4

Create a feasible region for the following linear programming problem and use the objective line method to find the vertex that gives the optimal solution.

Minimise $P = 2x + y$
subject to $3x + 4y \geqslant 24$
$$y \leqslant 3x$$
$$x \geqslant 0, y \geqslant 3.$$

Discussion point

➜ Explain why the constraint $y \leqslant 3x$ is equivalent to 'The ratio of y to x is no greater than 3:1, for $x \neq 0$.'

Solution

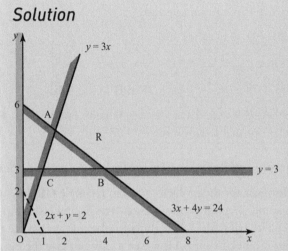

Figure 6.3

Figure 6.3 shows the feasible region, R.

In this case, the line $P = 2x + y$ needs to be parallel to $2x + y = 2$, but as near to the origin as possible.

The diagram shows that P is minimised at vertex A.

ACTIVITY 6.1

Find the optimal solution for the linear programming problem given in Example 6.4, assuming that non-integer solutions are allowed.

Example 6.5

A vet is treating a farm animal and she must provide a minimum daily amount of antibiotics, vitamins and nutrients in tablet or liquid form.

Table 6.1 shows what is needed and what is contained in the medicine.

	Antibiotic	Vitamin	Nutrient
Tablets (units per tablet)	3	2	10
Liquid (units per dose)	2	4	50
Daily requirement (units)	18	16	100

Table 6.1

(i) If the tablets cost 38 p each, and the liquid medicine costs £1 per dose, formulate this as a linear programming problem.

(ii) Solve the linear programming problem graphically.

(iii) The animal struggles to swallow many tablets so the vet decides that the number of doses of the liquid medicine should be at least double the number of tablets. How does this affect the solution?

(iv) The solution in (iii) is too expensive so the vet decides to reduce the daily requirement of antibiotics so that the cost is below £5. Find the new daily requirement and the new cost.

Solution

(i) Let x be the number of tablets taken per day, and y be the number of daily doses of the liquid medicine.

Minimise $0.38x + y$

Subject to: $3x + 2y \geqslant 18$

                $2x + 4y \geqslant 16$

          $10x + 50y \geqslant 100$

(ii)

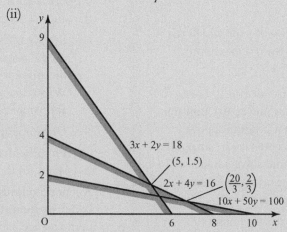

Figure 6.4

An integer solution is required so search close to the marked vertices of the feasible region for a point that satisfies the constraints and has the least cost:

£3.28 for 6 tablets and one dose

(iii) The additional constraint can be expressed as $y \geqslant 2x$.

The constraints $2x + 4y \geqslant 16$ and $10x + 50y \geqslant 100$ become redundant $y = 2x$ intersects with $3x + 2y = 18$ at $\left(\frac{18}{7}, \frac{36}{7}\right)$ or $\left(2\frac{4}{7}, 5\frac{1}{7}\right)$.

Examining the integer points close to $\left(2\frac{4}{7}, 5\frac{1}{7}\right)$, $(2, 6)$ satisfies both constraints with cost £6.76

Reducing x in favour of y increases the cost.

Reducing y in favour of x moves the solution out of the feasible region.

The solution is 2 tablets and 6 doses of liquid medicine at a cost of £6.76

(iv) By inspection, $(2, 4)$ satisfies both the vitamin and nutrient constraints. The number of doses of liquid medicine is double the number of tablets and the cost is £4.76

$3x + 2y = 3 \times 2 + 2 \times 4 = 14$ so 14 is the new daily requirement for the antibiotic.

Exercise 6.1

① Solve the following linear programming problem.

Maximise $\qquad P = x + 2y$

subject to $\qquad 4x + 5y \leqslant 45$

$\qquad\qquad\quad 4x + 11y \leqslant 44$

$\qquad\qquad\qquad x + y \leqslant 6.$

② A farmer grows two crops: wheat and beet. The number of hectares of wheat, x, and the number of hectares of beet, y, must satisfy

$10x + 3y \leqslant 52$

$\ \ 2x + 3y \leqslant 18$

$\qquad\ \ y \leqslant 4.$

Determine the values of x and y for which the profit function, $P = 7x + 8y$, is a maximum. State the maximum value of P.

③ A robot can walk at $1.5\,\mathrm{m\,s^{-1}}$ or run at $4\,\mathrm{m\,s^{-1}}$. When walking it consumes power at 1 unit per metre, and when running it consumes power at three times this rate. Its batteries are charged to 9000 units. Formulate and solve a linear programming problem to find the greatest distance it can cover in half an hour?

④ A builder can build either luxury houses or standard houses on a plot of land. Planning regulations prevent the builder from building more than 30 houses altogether, and he wants to build at least 5 luxury houses and at least 10 standard houses. Each luxury house requires $300\,\mathrm{m^2}$ of land, and each standard house requires $150\,\mathrm{m^2}$ of land. The total area of the plot is $6500\,\mathrm{m^2}$.

Given that the profit on a luxury house is £14 000 and the profit on a standard house is £9000, find how many of each type of house he should build to maximise his profit.

⑤ Maximise $\qquad z = x + y$

subject to $\qquad 3x + 4y \leqslant 12$

$\qquad\qquad\quad 2x + y \leqslant 4$

$\qquad\qquad\quad x \geqslant 0, y \geqslant 0$

$\qquad\qquad\quad x$ integer, y integer.

⑥ The Chief Executive of Leschester City Football Club plc has up to £4 million to spend following a good cup run. He has to decide on spending priorities.

Money needs to be spent on strengthening the playing squad and on extra support facilities (i.e. non-playing staff and stadium facilities).

The coach, who is popular with the fans, has said that he will resign unless he gets at least £2 million to spend on new players.

The authorities require that at least £0.6 million be spent to remedy stadium deficiencies affecting crowd safety.

Club policy is that the amount to be spent on support facilities must be at least one quarter of the amount to be spent on the playing squad.

(i) Let £x million be the amount to be spent on the playing squad and let £y million be the amount to be spent on support facilities. Write down four inequalities in terms of x and y representing constraints on spending.

(ii) Draw a graph to illustrate your inequalities.

(iii) Find the maximum amount that may be spent on the playing squad.

A report commissioned from a market research company indicates that fans regard both team performance and facilities as being important. The report states that the function $0.8x + 0.2y$ gives a measure of satisfaction with extra expenditure.

The Chief Executive proposes to spend £2.5 million on the playing squad and £1.5 million on support facilities.

(iv) Calculate the measure of satisfaction corresponding to the Chief Executive's proposals.

(v) Add to your graph the line $0.8x + 0.2y = 2.3$, and explain what points on this line represent.

(vi) The coach argues that the Chief Executive can achieve the same satisfaction score by spending less in total, but more on the playing squad. How much less and how much more?

[MEI]

⑦ Two products, X and Y, require three ingredients, A, B and C, for their manufacture. Table 6.2 summarises the amounts required and how much of each is available.

	Resource A	Resource B	Resource C
Amount required per unit of product — Product X	15	10	8
Amount required per unit of product — Product Y	5	7	12
Amount available	600	560	768

Table 6.2

It is required to maximise the total output of the two products subject to the amounts available.

(i) Identify variables and formulate an appropriate linear programming problem.

(ii) Solve your linear programming problem graphically, and interpret the solution.

(iii) The amount of B available is increased by 16. Show that the total output can be increased by 1 unit.

(iv) The amount of B available is increased by a further 16. Show that the total output cannot be increased any further. [MEI]

⑧ Coal arrives at a coal preparation plant from an opencast site and from a deep mine. It is to be mixed to produce a blend for an industrial customer. The customer requires 20 000 tonnes per week of the blend, and will pay £20 per tonne. Deep-mined coal has a marginal cost of £10 per tonne and coal from the opencast site has a marginal cost of £5 per tonne.

The blend must contain no more than 0.17% chlorine, since otherwise the hydrochloric acid produced by burning would corrode the boilers.

The blend must contain no more than 2% sulphur, since this burns to produce sulphur dioxide which subsequently dissolves to give acid rain. Acid rain damages the environment.

The blend must produce no more than 20% ash when burnt, otherwise the boilers will clog.

The blend must contain no more than 10% water, since otherwise the calorific value is affected.

The deep-mined coal has a chlorine content of 0.2%, a sulphur content of 3%, ash of 35% and water 5%. The opencast coal has a chlorine content of 0.1%, sulphur of 1%, ash of 10% and a water content of 12%.

Formulate and solve a linear programming problem to find how much of each type of coal should be blended to satisfy the contract with maximum profit? Which constraint is critical and which constraints are redundant?

⑨ A furniture manufacturer produces tables and chairs. A table requires £20 worth of materials and 10 person hours of work. It sells for a profit of £15.

Each chair requires £8 of materials and 6 hours of work. The profit on a chair is £7.

Given that £480 and 300 worker hours are available for the next production batch, find how many tables and chairs should be produced to maximise the profit.

Why might the optimal solution not be a practical solution?

⑩ A clothing retailer stocks two types of jacket which cost her £10 and £30 to purchase. She sells them at £20 and £50 respectively. She needs to order at least 200 jackets and has £2700 to spend.

The cheaper jackets need 20 cm of hanging space. The expensive jackets need only 10 cm each. She has 40 m of hanging space.

(i) Formulate a linear programming problem, assuming that all the jackets will be sold and that the retailer wishes to maximise her profit.

(ii) Solve the problem using a graphical method.

(iii) What would be the effect of trying to increase the order to satisfy a 10% increase in the demand for jackets? Explain your answer.

2 Branch-and-bound method

Where an integer solution is required (sometimes referred to as an **integer (linear) programming** problem), it has been shown how neighbouring grid points can be investigated. However, the solution obtained cannot be guaranteed to be optimal. The **branch-and-bound** method is a procedure for partitioning the feasible region and systematically examining each sub-region. It has applications in areas other than linear programming.

The procedure is potentially lengthy, but you are likely to be asked to solve only a simple case, such as those shown in the following examples. Any single example will not cover all the issues that may need to be addressed, but there is a summary of the issues involved on page 112.

Example 6.6

Maximise $P = 2x + 3y$

subject to $x + 2y \leqslant 12$

$3x + y \leqslant 15$

$y \leqslant 5$

$x \geqslant 0, y \geqslant 0$ (x and y integers).

ACTIVITY 6.2

Solve this problem, assuming that non-integer solutions are allowed.

ACTIVITY 6.3

Find an integer solution by investigating neighbouring grid points.

Note

Where the coefficients in the objective function are non-integer, the upper bound would be set to 19.8.

Note

Some versions of the branch and bound method avoid using the expressions lower and upper bounds.

Solution

The constraint and objective lines are shown in Figure 6.5.

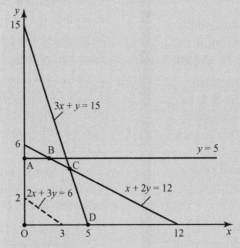

Figure 6.5

The branch-and-bound method can be represented by a tree diagram (see Figure 6.6), which is explained below.

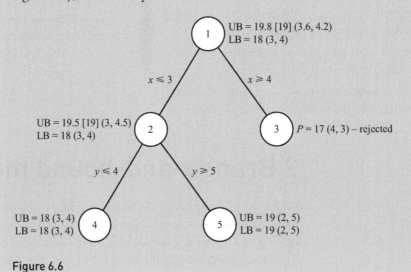

Figure 6.6

The original, non-integer ('**relaxed**') solution is recorded at node 1.

In this example, the maximum possible value for P with a non-integer solution is 19.8, but as the coefficients of x and y in the objective function ($P = 2x + 3y$) are integers, it follows that the upper bound for P amongst integer solutions is effectively 19.

To find a lower bound for P, consider the solution (3, 4), obtained by truncating 3.6 and 4.2. This gives $P = 18$. So it is known that the optimal value for P is at least 18.

There is scope for an improvement to this: the question is whether an integer solution exists where $P = 19$.

The next step in the process is to explore two branches from node 1: $x \leqslant 3$ and $x \geqslant 4$. Instead $y \leqslant 4$ and $y \geqslant 5$ could have been chosen, but it is conventional to select the variable with the biggest jump to its truncated value (i.e. 0.6 in this case), as this often produces a solution more quickly.

The branch $x \leqslant 3$ leads to node 2, whilst the branch $x \geqslant 4$ leads to node 3.

At node 2, the original problem is now solved, with the addition of the constraint $x \leqslant 3$.

From Figure 6.5 it can be seen that P will be maximised at the vertex of the (new) feasible region where the lines $x = 3$ and $x + 2y = 12$ intersect, giving the solution (3, 4.5), with $P = 19.5$.

This is not an integer solution, but in an example where a non-integer value for P was possible it would have been established that the upper bound was reduced to 19.5 for this node. This information is recorded at node 2, along with the fact that the lower bound is still 18.

Now consider node 3, which has the original constraints of the problem, with the addition of the constraint $x \geqslant 4$. From Figure 6.5 you can see that P will be maximised at the vertex of the (new) feasible region where the lines $x = 4$ and $3x + y = 15$ intersect, giving the solution (4, 3), with $P = 17$. So an integer solution has been found, but unfortunately the value for P is less then the lower bound of 18, and so can be rejected. This means that no further progress can be made from this node.

Observing that the y-value in (3, 4.5) is truncated to 4, the branches $y \leqslant 4$ and $y \geqslant 5$ from node 2 are explored, leading to nodes 4 and 5.

At node 4, the original problem can now be solved, with the addition of the constraint $y \leqslant 4$, as well as $x \leqslant 3$.

From Figure 6.5 it can be seen that P will be maximised at the vertex of the (new) feasible region where the lines $y = 4$ and $x = 3$ intersect, giving the solution (3, 4) with $P = 18$. This is the upper bound for node 4 (and all branches leading off it). As it is not an improvement on the existing lower bound, there is no need to investigate branches from node 4.

At node 5, the original problem can now be solved, with the addition of the constraint $y \geqslant 5$, as well as $x \leqslant 3$. As there is the additional constraint $y \leqslant 5$, this means that $y = 5$. From Figure 6.5 it can be seen that P will be maximised at the vertex of the (new) feasible region where the lines $y = 5$ and $x + 2y = 12$ intersect, giving the solution (2, 5) with $P = 19$. So an integer solution has been found with a value of P equal to the largest upper bound amongst the end nodes. For completeness, the improved lower bound of 19 at node 5 can be recorded.

Note

In some cases, a large number of nodes may need to be investigated. However, the upper bounds of the nodes will gradually decrease whilst the lower bound for the whole tree will gradually increase. Eventually, this will trap an integer solution, although it may just be the one associated with the initial lower bound.

Note

If the integer restriction only applies to one of the variables, then the above procedure can still be carried out, with natural modifications. Thus, only the restricted variable is rounded down when finding the lower bound, and only the restricted variable is branched on.

In Example 6.6, it was clear which branch needed to be examined at each stage. Had the solution at node 3 not been rejected, it would have been best to consider the branch from the node with the largest upper bound. For example, if the upper bound at node 3 had been 19.6, in a situation where non-integer values for p were possible, then branches from node 3 would have been investigated before those from node 2.

Summary of the branch-and-bound method (for maximisation)

- Record the non-integer solution at node 1, together with the resulting value of the objective function, which is the upper bound (often restricted to an integer).

- Truncate the variables to obtain an integer solution, giving a lower bound. Record this at node 1 as well.

- Branch on the variable that was truncated the most (to give, for example, $x \leq 3$ and $x \geq 4$ as the two branches).

- Solve the original problem again, with the added constraint $x \leq 3$ (for example), recording the solution at the node in question, together with the (possibly reduced) upper bound (which applies to all branches from that node).

- If the solution obtained is a purely integer one, it may produce an increase in the lower bound for the tree. The current lower bound is recorded at each node.

- When choosing the next node to branch from, select the one with the largest upper bound. The two branches are obtained by truncating the non-integer variable.

- When adding constraints to the original problem, include those on the earlier branches.

- Nodes are rejected when the value of the objective function is less than the current lower bound.

- An optimal integer solution is obtained if the value of the objective function is equal to the largest upper bound amongst the remaining end nodes.

If the objective function has to be minimised, then the process is the same, except that the roles of the upper and lower bounds are reversed, as shown in the next example.

Example 6.7

Minimise $P = 3x + 2y$

subject to $2x + y \geq 10$

$5x + 8y \geq 40$

$x \geq 0, y \geq 0$ (x and y integers).

> **Note**
>
> For a minimisation problem, the lower bound for each node is recorded, together with the upper bound inherited from earlier nodes (representing the best integer solution found so far).

Solution

The constraint and objective lines are shown in Figure 6.7.

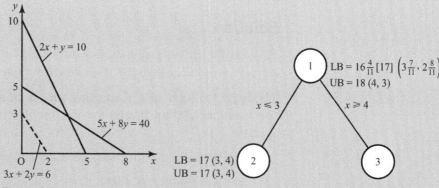

Figure 6.7

Figure 6.8

Referring to Figures 6.7 and 6.8, the relaxed solution occurs at the vertex of the feasible region where $2x + y = 10$ and $5x + 8y = 40$, giving $\left(3\frac{7}{11}, 2\frac{8}{11}\right)$ where $P = 16\frac{4}{11}$.

So at node 1 the lower bound is $16\frac{4}{11}$ (though, once again, it is effectively 17, because of the integer coefficients in the objective function).

In order to find the upper bound, the values of $3\frac{7}{11}$ and $2\frac{8}{11}$ are rounded up to give $(4, 3)$, where $P = 18$.

At this point, consider the branches $x \leqslant 3$ and $x \geqslant 4$ (by convention, as the distance from $3\frac{7}{11}$ to 4 is greater than that from $2\frac{8}{11}$ to 3).

At node 2, the original problem can be solved, with the addition of the constraint $x \leqslant 3$. From Figure 6.7 it can be seen that P will be maximised at the vertex of the (new) feasible region where the lines $x = 3$ and $2x + y = 10$ intersect, giving the solution $(3, 4)$ with $P = 17$. So an integer solution has been found with a value of P equal to the effective lower bound. For completeness, the improved upper bound of 17 at node 2 can be recorded. Note that there is no need to investigate node 3.

Exercise 6.2

For the following questions, use the branch-and-bound method to find optimal integer solutions.

① Maximise $P = x + 2y$
subject to $4x + 11y \leqslant 44$
$x + y \leqslant 6$
$x \geqslant 0, y \geqslant 0$

② Maximise $P = 7x + 8y$
subject to $10x + 3y \leqslant 52$
$2x + 3y \leqslant 18$
$y \leqslant 4$
$x \geqslant 0, y \geqslant 0$

③ Minimise $P = 3x + 7y$
subject to $12x + 5y \geqslant 60$
$y \leqslant 12x$
$y \geqslant 6$
$x \geqslant 0$

3 The simplex algorithm

The simplex algorithm provides an algebraic method for dealing with linear programming problems, suitable for use by a computer. It can also deal with problems involving more than two variables, which can't be represented graphically.

Example 6.8

Apply the simplex algorithm to the linear programming problem from Example 6.1.

> This problem involves two variables only.

Solution

First convert the inequalities into equations. $2x + 3y \leqslant 12$ becomes $2x + 3y + s = 12$, where $s \geqslant 0$ is known as a **slack variable**.

> Slack variables are needed to turn inequalities into equations

The larger the value of s_1, the further you are from the constraint line, and the greater the amount of slack available.

Including the objective function, the equations are

$$
\begin{aligned}
P - x - y &= 0 \quad \text{①} \\
2x + 3y + s &= 12 \quad \text{②} \\
6x + 5y + t &= 30 \quad \text{③} \qquad (s, t \geqslant 0)
\end{aligned}
$$

These equations are set up so that all the variables appear on the left-hand side. The equations can also be presented using the **simplex tableau**, as shown in Table 6.3.

> **Note**
>
> This is the expected form of presentation, but can be awkward to read to start with. The method will be explained using the equations.

P	x	y	s	t	RHS	Equation
1	−1	−1	0	0	0	①
0	2	3	1	0	12	②
0	6	5	0	1	30	③

Table 6.3

Start with an initial 'solution' of $x = 0$, $y = 0$. This satisfies the equations above, and gives $P = 0$, $s = 12$, $t = 30$.

It corresponds to O in the feasible region in Figure 6.2.

> **Note**
>
> Here it was chosen to set $x = 0$, but you could have chosen y, or any variable with a negative coefficient in the equation of the objective function (including, in fact, slack variables).

To find a better solution, that gives a larger value for P, you can (for example) set $x = 0$, so that equations ① – ③ become

$$P = y, 3y + s = 12 \text{ and } 5y + t = 30.$$

This corresponds to working along the y-axis.

You now want to make y as large as possible. Noting the restriction that s and t have to be $\geqslant 0$, it will be possible to set y equal to the lower of $\frac{12}{3}$ and $\frac{30}{5}$: i.e. 4.

This step is referred to as the **ratio test**.

> The ratio test ensures that you are working within the feasible region.

This gives $P = 4$, $x = 0$, $y = 4$, $s = 0$ and $t = 10$, which corresponds to vertex A of the feasible region. This is a feature of the Simplex method, where there are just two control variables. You are working your way round the vertices of the feasible region (these vertices are sometimes referred to as **basic feasible solutions**) until no further improvement can be made.

Where the equation for the objective row is $P - 3x - 2y = 5$, for example, it is conventional to aim to maximise x (setting $y = 0$), because this gives the best chance of maximising P, on the grounds that, out of a choice of $5 + 3x$ and $5 + 2y$, $5 + 3x$ is more likely to be the larger (though this may not prove to be the case).

Continuing with y as the variable to be maximised, the y column in the Simplex tableau is called the **pivot column**, and the row that provides the maximum value of y from the ratio test is called the **pivot row**. In this example, this is the row corresponding to equation ②.

The pivot row is used to consolidate the improvement in P, as follows.

The aim is to eliminate y from all of the rows except the pivot row. You use a method that a computer can easily apply. You first divide both sides of equation ② by 3, so that the coefficient of y becomes 1. Equation ② is now

$$\frac{2x}{3} + y + \frac{s}{3} = 4 \quad ②a$$

Now eliminate y from equation ①, by adding equation ②a to equation ①, to give

$$P - \frac{x}{3} + \frac{s}{3} = 4.$$

Notice that, by setting x and s equal to zero, you can obtain the improved value of $P = 4$.

You also eliminate y from equation ③, by subtracting 5 times equation ②a, to give

$$\frac{8x}{3} - \frac{5s}{3} + t = 10.$$

To summarise, the new equations are (relabelling ②a as ⑤)

$$P - \frac{x}{3} + \frac{s}{3} = 4 \qquad ④ = ① + ⑤$$
$$\frac{2x}{3} + y + \frac{s}{3} = 4 \qquad ⑤ = ② \div 3$$
$$\frac{8x}{3} - \frac{5s}{3} + t = 10 \qquad ⑥ = ③ - 5 \times ⑤.$$

The simplex tableau is now as follows.

P	x	y	s	t	RHS	Equation
1	$-\frac{1}{3}$	0	$\frac{1}{3}$	0	4	④
0	$\frac{2}{3}$	1	$\frac{1}{3}$	0	4	⑤
0	$\frac{8}{3}$	0	$-\frac{5}{3}$	1	10	⑥

Table 6.4

Note that y now appears only once in the pivot column, with a coefficient of 1. Any value chosen for y will not affect the rows in which this variable has a zero coefficient. Similarly, for P and t. These variables are termed **basic** variables, whilst the variables that have coefficients in more than one row are termed **non-basic**.

At the end of each stage of the simplex method, you can obtain the (current) improved solution by setting the non-basic variables to zero. This means that the values of the basic variables will be those in the right-hand column.

So $P = 4$, $x = 0$, $y = 4$, $s = 0$ and $t = 10$.

This corresponds to point A on the graphical representation.

Example 6.9

Apply a second iteration of the simplex algorithm, to obtain an improved value for P.

Solution

Take x as the pivot column, and apply the ratio test. You see that equation ⑤ gives $\frac{4}{\left(\frac{2}{3}\right)} = 6$, whilst equation ⑥ gives $\frac{10}{\left(\frac{8}{3}\right)} = \frac{15}{4} < 6$.

> Always show your working for the ratio test and write down the pivot row.

So equation ⑥ is the pivot row.

Ensure that the coefficient of x is 1, then obtain new equations by eliminating x from ④ and ⑤, giving:

$$P + \frac{s}{8} + \frac{t}{8} = \frac{21}{4} \qquad ⑦ = ④ + \frac{1}{3} \times ⑨$$

$$y + \frac{3s}{4} - \frac{t}{4} = \frac{3}{2} \qquad ⑧ = ⑤ - \frac{2}{3} \times ⑨$$

$$x - \frac{5s}{8} + \frac{3t}{8} = \frac{15}{4} \qquad ⑨ = \frac{3}{8} \times ⑥$$

The simplex tableau is now as follows.

P	x	y	s	t	RHS	Equation
1	0	0	$\frac{1}{8}$	$\frac{1}{8}$	$\frac{21}{4}$	⑦
0	0	1	$\frac{3}{4}$	$-\frac{1}{4}$	$\frac{3}{2}$	⑧
0	1	0	$-\frac{5}{8}$	$\frac{3}{8}$	$\frac{15}{4}$	⑨

Table 6.5

Now P, x and y are the basic variables, and the improved solution is $P = \frac{21}{4}, x = \frac{15}{4}$

> Note that the value for y has reduced from 4 to $\frac{3}{2}$: i.e. the maximum value wasn't needed in the end.

$y = \frac{3}{2}, s = 0$ and $t = 0$,

which corresponds to vertex B of the feasible region.

As the coefficients of s and t in the objective row are both positive, there is no further scope for increasing P.

In the previous example, the aim was to maximise P. If, instead, you wish to minimise P, then this can sometimes be done by maximising $-P$ as in the following example. This also features three control variables (and so couldn't be solved by a graphical method). At this point, a couple of restrictions on the simplex method should be mentioned. Firstly, inequalities need to be of the form $2x + 3y \leqslant 12$, rather than $2x + 3y \geqslant 12$. This ensures that the initial solution $x = 0, y = 0$ satisfies the constraints. Secondly, when the equations are set up for the constraints, the values on the right-hand side need to be non-negative. This ensures that the ratio test can be carried out in the same way each time. This doesn't apply to the objective row.

Example 6.10

Minimise $2x + y - 3z$, subject to the following constraints.

$x - 4y + z \leqslant 4$

$3x + 2y - z \geqslant -2$

$x \geqslant 0, y \geqslant 0, z \geqslant 0$

Solution

Step 1 Rewrite the problem as

Maximise $P = -2x - y + 3z$

subject to $x - 4y + z \leqslant 4$

and $-3x - 2y + z \leqslant 2$.

> Note that the inequalities are now of the $\leqslant$ type, and that the right-hand values are non-negative.

Step 2 Create equations, with slack variables.

$$
\begin{aligned}
P + 2x + y - 3z & = 0 \quad \text{①} \\
x - 4y + z + s & = 4 \quad \text{②} \\
-3x - 2y + z + t & = 2 \quad \text{③}
\end{aligned}
$$

Step 3 Represent the equations in a simplex tableau.

P	x	y	z	s	t	RHS	Equation
1	2	1	−3	0	0	0	①
0	1	−4	1	1	0	4	②
0	−3	−2	1	0	1	2	③

Table 6.6

Step 4 Choose z as the pivot column (as it is the only variable in the objective row with a negative coefficient), and apply the ratio test.

② with $x = y = s = 0 \Rightarrow z = 4$

③ with $x = y = t = 0 \Rightarrow z = 2$

As $2 < 4$, ③ is the pivot row (indicated by circling the coefficient of z in equation 3 in the table below).

P	x	y	z	s	t	RHS	Equation
1	2	1	−3	0	0	0	①
0	1	−4	1	1	0	4	②
0	−3	−2	①	0	1	2	③

Table 6.7

Step 5 Eliminate z from equations 1 and 2.

P	x	y	z	s	t	RHS	Equation
1	−7	−5	0	0	3	6	④ = ① + 3 × ⑥
0	4	−2	0	1	−1	2	⑤ = ② − ⑥
0	−3	−2	1	0	1	2	⑥ = ③

Table 6.8

Step 6 As x now has the largest negative coefficient in the new objective row (equation ④), you choose this as the next pivot column.

You would normally apply the ratio test now, but notice that the coefficient of x in equation ⑥ is negative. As explained in the following note (which you may wish to omit on a first reading), any such row can be excluded when establishing the pivot row.

Note

- -

The ratio test

In Step 6 of Example 6.10, you are trying to maximise x, so that when y and t are set equal to zero in equation ④, P (equal to $6 + 7x$) will be increased by as much as possible.

When maximising x you need to ensure that all the constraint equations are satisfied. For equation ⑤, you set s equal to zero (you have already set y and t to zero in equation ④), to leave you with $4x = 2$. This places an upper limit of $\frac{2}{4} = \frac{1}{2}$ on x (x could be less than $\frac{1}{2}$, as it would always be possible to introduce another slack variable, giving $4x + u = 2$).

In equation ⑥, the coefficient of x is negative, and this row cannot be chosen as the pivot row: such a choice would require you to subtract a multiple of the pivot row from the objective row (in order to eliminate x from it), and this would cause the value of the objective row to fall. This assumes that the values in the right-hand column of the constraint rows are non-negative (which is a requirement of the method). So, when applying the ratio test, you only need to consider constraint equations where the coefficient of the pivot variable is positive.

So you take the pivot row to be equation ⑤, and divide by 4 before eliminating x from the other equations, to give equations ⑦–⑨.

P	x	y	z	s	t	RHS	Equation
1	−7	−5	0	0	3	6	④ = ① + 3 × ⑥
0	④	−2	0	1	−1	2	⑤ = ② − ⑥
0	−3	−2	1	0	1	2	⑥ = ③

Table 6.9

P	x	y	z	s	t	RHS	Equation
1	0	$-8\frac{1}{2}$	0	$1\frac{3}{4}$	$1\frac{1}{4}$	$9\frac{1}{2}$	⑦ = ④ + 7 × ⑧
0	1	$-\frac{1}{2}$	0	$\frac{1}{4}$	$-\frac{1}{4}$	$\frac{1}{2}$	⑧ = ⑤ ÷ 4
0	0	$-3\frac{1}{2}$	1	$\frac{3}{4}$	$\frac{1}{4}$	$3\frac{1}{2}$	⑨ = ⑥ + 3 × ⑧

Table 6.10

Step 7 There are no further variables that have negative coefficients in the objective row and positive coefficients elsewhere in their columns, so you have reached the end of the process.

The variables with a single 1 (and otherwise zeros) in their columns are the basic variables, and their values can be read from the right-hand column, whilst the other (non-basic) variables are set to zero.

Solution: $x = 0.5, y = 0, z = 3.5, s = 0, t = 0$.

The maximised value of $-2x - y + 3z$ is 9.5 and so the minimised value of $2x + y - 3z$ is -9.5.

Step 8 Check:

$x - 4y + z = 4$ $(\leqslant 4)$ and $3x + 2y - z = -2$ $(\geqslant -2)$.

Exercise 6.3

① Use the simplex algorithm to solve the following linear programming problem.

Maximise $P = 9x + 10y + 6z$

subject to $2x + 3y + 4z \leqslant 3$

$\qquad 6x + 6y + 2z \leqslant 8$

$\qquad x \geqslant 0, y \geqslant 0, z \geqslant 0.$

② Use the simplex algorithm to solve the following linear programming problem.

Maximise $P = 3w + 2x$

subject to $w + x + y + z \leqslant 150$

$\qquad 2w + x + 3y + 4z \leqslant 200$

$\qquad w \geqslant 0, x \geqslant 0, \ y \geqslant 0, z \geqslant 0.$

③ Use the simplex algorithm to solve the following linear programming problem.

Maximise $P = 3w + 2x$

subject to $w + x + y + z \leqslant 150$

$\qquad 2w + x + 3y + 4z \leqslant 200$

$\qquad w \geqslant x$ (rewrite this as $x - w \leqslant 0$)

$\qquad w \geqslant 0, x \geqslant 0, y \geqslant 0, z \geqslant 0.$

④ The 'Cuddly Friends Company' produces soft toys. For one day's production run it has available $11\,m^2$ of furry material, $24\,m^2$ of woolly material and 30 glass eyes. It has three soft toys that it can produce.

The 'Cuddly Aardvark', each of which requires $0.5\,m^2$ of furry material, $2\,m^2$ of woolly material and two eyes. Each sells at a profit of £3.

The 'Cuddly Bear', each of which requires $1\,m^2$ of furry material, $1.5\,m^2$ of woolly material and two eyes. Each sells at a profit of £5.

The 'Cuddly Cat', each of which requires $1\,m^2$ of furry material, $1\,m^2$ of woolly material and two eyes. Each sells at a profit of £2.

An analyst formulates the following linear programming problem to find the production plan that maximises profit.

Maximise $\qquad 3a + 5b + 2c$

subject to $\qquad 0.5a + b + c \leqslant 11$

$\qquad\qquad 2a + 1.5b + c \leqslant 24$

$\qquad\qquad 2a + 2b + 2c \leqslant 30$

(i) Explain how this formulation models the problem, and say why the analyst has not simplified the last inequality to $a + b + c \leqslant 15$.

(ii) The final constraint is different from the others in that the resource is integer valued. Explain why that does not impose an additional difficulty for this problem.

(iii) Solve this problem using the Simplex algorithm.

Interpret your solution and say what resources are left over. [MEI adapted]

⑤ A publisher is considering producing three books over the next week: a mathematics book, a novel and a biography. The mathematics book will sell at £10 and costs £4 to produce. The novel will sell at £5 and costs £2 to produce. The biography will sell at £12 and costs £5 to produce. The publisher wants to maximise profit, and is confident that all books will be sold.

There are constraints on production. Each copy of the mathematics book needs 2 minutes of printing time, 1 minute of packing time, and $300\,cm^3$ of temporary storage space.

Each copy of the novel needs 1.5 minutes of printing time, 0.5 minutes of packing time, and $200\,cm^3$ of temporary storage space.

Each copy of the biography needs 2.5 minutes of printing time, 1.5 minutes of packing time, and $400\,cm^3$ of temporary storage space.

There are 10 000 minutes of printing time available on several printing presses, 7500 minutes of packing time, and $2\,m^3$ of temporary storage space.

(i) Explain how the following initial feasible tableau models this problem.

P	x	y	z	s	t	u	RHS
1	−6	−3	−7	0	0	0	0
0	2	1.5	2.5	1	0	0	10 000
0	1	0.5	1.5	0	1	0	7 500
0	300	200	400	0	0	1	2 000 000

Table 6.11

(ii) Use the simplex algorithm to solve your linear programming problem, and interpret your solution.

(iii) The optimal solution involves producing just one of the three books. By how much would the price of each of the other books have to be increased to make them worth producing? [MEI adapted]

Exercise 6.4

① A factory's output includes three products. To manufacture a tonne of product A, 3 tonnes of water are needed. Product B needs 2 tonnes of water per tonne produced, and product C needs 5 tonnes of water per tonne produced.

Product A uses 5 hours of labour per tonne produced, product B uses 6 hours and product C uses 2 hours.

There are 60 tonnes of water and 50 hours of labour available for tomorrow's production.

(i) Formulate a linear programming problem to maximise production within the given constraints.

(ii) Use the simplex algorithm to solve your linear programming problem, **choosing the first pivot to make c a basic variable**.

[MEI adapted]

② A farmer has 40 acres of land. Four crops, A, B, C and D are available.

Crop A will return a profit of £50 per acre.
Crop B will return a profit of £40 per acre.
Crop C will return a profit of £40 per acre.
Crop D will return a profit of £30 per acre.
The total number of acres used for crops A and B must not be greater than the total number used for crops C and D.

The farmer formulates this problem as:

Maximise $50a + 40b + 40c + 30d$,

subject to $a + b \leqslant 20$

$a + b + c + d \leqslant 40$

(i) Explain what the variables a, b, c and d represent.

Explain how the first inequality was obtained.

Explain why expressing the constraint on the total area of land as an inequality will lead to a solution in which all of the land is used.

(ii) Solve the problem using the simplex algorithm. [MEI]

③ A production unit makes two types of product, X and Y. Production levels are constrained by the availability of finance, staff and storage space. Requirements for, and daily availabilities of, each of these resources are summarised in Table 6.12.

	Finance (£)	Staff time (hours)	Storage space (m³)
Requirement per tonne for product X	400	8	1
Requirement per tonne for product Y	200	8	3
Resources available per day	2000	48	15

Table 6.12

The profit on these products is £320 per tonne for X and £240 per tonne for Y.

(i) Formulate the constraints for this problem.

(ii) State the objective function, assuming that profit is to be maximised.

(iii) Use a graphical method to find the best daily production plan.

(iv) Set up an initial tableau for the problem and use the simplex algorithm to solve it. Relate each stage of the tableau to its corresponding point on your graph.

④ In an executive initiative course, participants are asked to travel as far as possible in three hours using a combination of moped, car and lorry. The moped can be carried in the car and the car can be carried on the lorry.

The moped travels at 20 miles per hour (mph) with a petrol consumption of 60 miles per gallon (mpg). The car travels at 40 mph with a petrol consumption of 40 mpg. The lorry travels at 30 mph with a petrol consumption of 20 mpg.

2.5 gallons of petrol are available.

The moped must not be used for more than 55 miles, and a total of no more than 55 miles must be covered using the car and/or lorry.

Formulate this problem as a standard LP problem in terms of the two variables m = time spent travelling by moped (hours) and c = time spent travelling by car (hours).

Find the optimal solution and interpret it in the context of the problem.

⑤ A craft workshop produces three products: xylophones, yodellers and zithers. The times taken to make them and the total time available per week are shown in Table 6.13. Also shown are the costs and the total weekly capital available.

	Xylophones	Yodellers	Zithers	Resource availability
Time (hours)	2	5	3	30
Cost (£00s)	4	1	2	24

Table 6.13

Profits are £180 per xylophone, £90 per yodeller and £110 per zither.

(i)   Formulate a linear programming problem to find the weekly production plan that maximises profit within the resource constraints.

(ii) Use the simplex algorithm to solve the problem, pivoting first on the column of your tableau containing the variable that represents the number of xylophones produced. Explain how your final tableau shows that the workshop should produce 5 xylophones and 4 yodellers.

If, when applying the simplex algorithm, the first pivot is on the column containing the variable which represents the number of zithers produced, then the final solution obtained is for the workshop to produce 1.5 xylophones and 9 zithers per week.

(iii) How can this production plan be implemented?

(iv) Explain how the simplex algorithm can lead to different solutions.   [MEI adapted]

⑥ A furniture manufacturer is planning a production run. He will be making wardrobes, drawer units and desks. All can be manufactured from the same wood.

He has available 200 m² of wood for the production run. Allowing for wastage, a wardrobe requires 5 m², a drawer unit requires 3 m², and a desk requires 2 m².

He has 200 hours available for the production run. A wardrobe requires 4.5 hours. a drawer unit requires 5.2 hours, and a desk requires 3.8 hours.

The completed furniture will have to be stored at the factory for a short while before being shipped. The factory has 50 m³ of storage space available. A wardrobe needs 1 m³, a drawer unit needs 0.75 m³, and a desk needs 0.5 m³.

The manufacturer needs to know what he should produce to maximise his income. He sells

the wardrobes at £80 each, the drawer units at £65 each and the desks at £50 each.

(i)   Formulate the manufacturer's problem as a linear programming problem.

(ii) Use the simplex algorithm to solve the linear programming problem.

(iii) Interpret the results.   [MEI]

⑦ Three chemical products, X, Y and Z, are to be made. Product X will sell at 40p per litre and costs 30p per litre to produce. Product Y will sell at 40p per litre and costs 30p per litre to produce. Product Z will sell at 40p per litre and costs 20p per litre to produce. Three additives are used in each product. Product X uses 5 g per litre of additive A, 2 g per litre of additive B and 8 g per litre of additive C. Product Y uses 2 g per litre of additive A, 4 g per litre of additive B and 3 g per litre of additive C. Product Z uses 10 g per litre of additive A, 5 g per litre of additive B and 5 g per litre of additive C. There are 10 kg of additive A available, 12 kg of additive B, and 8 kg of additive C.

(i)   Explain how the initial feasible tableau shown below models this problem.

P	x	y	z	s	t	u	RHS
1	−10	−10	−20	0	0	0	0
0	5	2	10	1	0	0	10 000
0	2	4	5	0	1	0	12 000
0	8	3	5	0	0	1	8 000

Table 6.14

(ii) Use the simplex algorithm to solve your linear programming problem, and interpret your solution.

(iii) The optimal solution involves making two of the three products. By how much would the cost of making the third product have to fall to make it worth producing, assuming that the selling price is not changed?

⑧ Neil is refurbishing a listed building. There are two types of paint that he can use for the inside walls. One costs £1.45 per m² and the other costs £0.95 per m². He must paint the lower half of each wall in the more expensive paint. He has 350 m² of wall to paint. He has a budget of £400 for wall paint.

The more expensive paint is easier to use, and so Neil wants to use as much of it as possible.

Initially, the following linear programming problem is constructed to help Neil with his purchasing of paint.

Let x be the number of m² of wall painted with the more expensive paint.

Let y be the number of m² of wall painted with the less expensive paint.

Maximise $P = x + y$

subject to $1.45x + 0.95y \leqslant 400$

$y - x \leqslant 0$

$x \geqslant 0, y \geqslant 0$

(i)   Explain the purpose of the inequality $y - x \leqslant 0$.

(ii)   The formulation does not include the inequality $x + y \geqslant 350$. State what this constraint models and why it has been omitted from the formulation.

(iii)   Use the simplex algorithm to solve the linear programming problem, pivoting first on the '1' in the y column. Interpret your solution.

The solution shows that Neil needs to buy more paint. He negotiates an increase in his budget to £450.

(iv)   Find the solution to the linear programming problem given by changing $1.45x + 0.95y \leqslant 400$ to $1.45x + 0.95y \leqslant 450$, and interpret your solution.

Neil realises that, although he now has a solution, that solution is not the best for his requirements.

(v)   Explain why the revised solution is not optimal for Neil.   [MEI]

LEARNING OUTCOMES

Now you have finished this chapter, you should be able to

➤ formulate constrained optimisation problems

➤ solve constrained optimisation problems via graphical methods

➤ apply the branch-and-bound method for integer programming

➤ use the simplex algorithm for maximising an objective function, including the use of slack variables

➤ interpret a simplex tableau.

KEY POINTS

1   Linear programming is concerned with solving logistical problems involving linear constraints – constrained optimisation problems.

2   The first step is to formulate the constraints mathematically.

3   The objective function is the item that you wish to maximise (or minimise).

4   The feasible region is the region of the graph where all the constraints are satisfied.

5   The objective function is generally maximised (or minimised) at one of the vertices of the feasible region.

6   Neighbouring points may need to be considered if an integer-valued solution is required.

7   Alternatively, the branch-and-bound method can be used to find an optimal integer solution. See the summary in the text for the key points of this method.

8   The simplex algorithm provides an algebraic method for dealing with linear programming problems.

   ■ Inequalities are converted into equations using slack variables.

   ■ A simplex tableau is used to record the result of each iteration.

   ■ For each iteration of the process, a pivot column is established, and the ratio test is used to identify the pivot row.

   ■ After each iteration, the values of any basic variables can be read from the tableau, and the non-basic variables are set to zero.

Game theory

→ **Matching pennies** is the name for a simple game used in **game theory**. It is played between two players, Even and Odd. Each player has a penny and must secretly turn the penny to heads or tails. The players then reveal their choices simultaneously. If the pennies match (both heads or both tails), then Even keeps both pennies, so wins one from Odd (+1 for Even, −1 for Odd). If the pennies do not match (one heads and one tails), Odd keeps both pennies, so receives one from Even (−1 for Even, +1 for Odd).

Game theory aims to model situations involving competitors (or players), where each player's gain (or loss) depends not only on their decisions, but also the decisions of their competitors. It can be applied in areas such as business negotiations, election campaigns, and warfare.

This chapter starts with **zero-sum** games between two players. This means that if player 1 wins a certain amount, then player 2 loses that amount.

1 Pay-off matrices

Zero sum games

Suppose that the players have two options: **A and B**. The amount (in £) won by player 1, and lost by player 2, is given by the **pay-off matrix** in Table 7.1.

	Player 2 plays A	Player 2 plays B
Player 1 plays A	1	0
Player 1 plays B	−5	10

Table 7.1

> The pay-off matrix shows player 1's gains and player 2's losses.

The elements in the pay-off matrix are given from player 1's point of view (this is the standard convention).

If player 1 wants to have a chance of winning £10, they must choose option **B**, and accept the risk of losing £5 if player 2 chooses option **A**. The safer course of action is to choose option **A** and either win £1 or break even.

Because this is a zero-sum game, to consider things from player 2's point of view, you transpose the matrix and reverse the signs of the elements in the matrix, to give the matrix in Table 7.2.

> The matrix is transposed (i.e. rows become columns and vice versa) so that the rows now give player 2's gains and the columns give player 1's losses.

	Player 1 plays A	Player 1 plays B
Player 2 plays A	−1	5
Player 2 plays B	0	−10

Table 7.2 Player 2's gains

> The pay-off matrix now shows player 2's gains and player 1's losses. The values are the same but the signs are reversed because it is a zero-sum game.

Option **A** offers player 2 the chance of winning £5, with the risk of losing £1, and choosing option **B** means either breaking even or losing £10.

The choice that each player will make will depend on the following factors:

- Their knowledge of the pay-off matrix.
- Whether the choice of the other player can be predicted. If the game is repeated, the other player may always make the same choice.
- How risk averse is the player? They may be able to afford to lose a small sum of money, but not a large amount. If the game is to be repeated many times, a player may be able to bear an occasional moderate loss.

Predicting the other player's choice is of particular interest in game theory. For the current example, player 1 may conclude that player 2 is more likely to choose option **A**, and so player 1 may decide to choose option **A** themselves, and win £1 (rather than lose £5). Player 2 will then lose £1.

If player 2 discovers player 1's intention to choose option **A**, then it is better for player 2 to choose option **B**, and so break even.

Play-safe strategies

A **play-safe strategy** is where a player chooses the option with the best worst outcome. In other words, they assume that the other player penalises them the most by their choice.

In the previous example, if both players adopt a play-safe strategy, player 1 chooses option A (which has a worst outcome of £0, compared with the worst outcome of −£5 for option B), and player 2 also chooses option A (which has a worst outcome of −£1, compared with the worst outcome of −£10 for option B).

The result is that player 1 wins £1, and player 2 loses £1. It is assumed that both players have to play the game.

Now consider the pay-off matrix in Table 7.3, where player 1 has two choices and player 2 has three.

	Player 2 plays A	**Player 2 plays B**	**Player 2 plays C**
Player 1 plays A	2	−1	1
Player 1 plays B	1	−2	−3

Table 7.3

As usual, these are the pay-offs for player 1. The pay-offs for player 2 are shown in Table 7.4.

	Player 1 plays A	**Player 1 plays B**
Player 2 plays A	−2	−1
Player 2 plays B	1	2
Player 2 plays C	−1	3

Table 7.4 (player 2's gains)

The matrix has been transposed and the signs have been reversed.

The worst outcomes for player 1 can be shown in an additional column, as in Table 7.5 with the better option (from player 1's point of view) circled.

	Player 2 plays A	Player 2 plays B	Player 2 plays C	Worst
Player 1 plays A	2	−1	1	(−1)
Player 1 plays B	1	−2	−3	−3

Table 7.5

The worst outcomes for player 2 can also be shown in an additional column, as in Table 7.6, where the best option (from player 2's point of view) is circled.

	Player 1 plays A	Player 1 plays B	Worst
Player 2 plays A	−2	−1	−2
Player 2 plays B	1	2	(1)
Player 2 plays C	−1	3	−1

Table 7.6 (player 2's gains)

The play-safe strategy for player 1 is option A, whilst for player 2 it is option B.

In this case, if both players adopt the play-safe strategy, player 1 will not want to change their choice (assuming that player 2 doesn't change from option B), as −1 is better than −2.

Similarly, player 2 will not want to change their choice (assuming that player 1 doesn't change from option A), as 1 is better than both −2 and −1.

This is known as a **stable solution**. It is also called a **saddle point**, and the game is said to be in equilibrium. The stable solution exists because, in player 1's matrix, −1 is below both 2 and 1 (in row A), whilst −1 is above −2 (in column B).

So, where a stable solution exists, it is in both players' interests for them to adopt the play-safe strategy. The **value of the game** is defined as player 1's pay-off.

Although you considered player 2's matrix in the previous discussion, the standard procedure for identifying a stable solution is based on player 1's matrix.

You establish the worst outcomes for player 1, and place these in a column to the right of the pay-off matrix. These are called the **row minima** (i.e. the column shows the minimum for each row).

Because the elements of the matrix are now player 2's outcomes with the sign reversed, the column of worst outcomes for player 2 now becomes the row of maximum values for each of the columns. These are called the **column maxima**.

This new system of labelling is shown in Table 7.7.

	A	B	C	Row minima
A	2	−1	1	(−1)
B	1	−2	−3	−3
Column maxima	2	(−1)	1	

Table 7.7

This is the min (column maxima) or column minimax.

This is the max (row minima) or row maximin.

A stable solution occurs because −1 is the minimum value in row A and also the maximum value in column B.

In general, a stable solution occurs when the maximum of the row minima equals the minimum of the column maxima.

Note

In 3D geometry, a point on a surface is a saddle point if it is a minimum point of the surface in one direction, but a maximum in another.

Note

So far, you have used an additional matrix from player 2's point of view. It is conventional to work only with the matrix from player 1's point of view, and this will be done from now on.

Unless stated otherwise, this means that the matrices will typically just have A, B, etc. in the column headings rather than the name of the strategy, 'Player 2 plays A', etc. The same goes for the 'Player 1 plays ...' entries in the row headings.

A zero-sum game has a stable solution if and only if

the max (row minima) = the min (column maxima)

If a stable solution exists, it is the best strategy for both players.

Converting from a constant-sum game to a zero-sum game

Many games are not zero-sum games. Some have a constant sum: these can be converted to zero-sum games. It is then possible to use the techniques for zero-sum games on them.

Discussion point

→ How could you find a play-safe strategy for the game in Example 7.1.

Example 7.1

Consider the following pay–off matrix (the numbers could represent market shares of two companies).

	A	B
A	60, 40	50, 50
B	30, 70	80, 20

Table 7.8

This means that, for example, if player 1 chooses B and player 2 chooses A, then player 1 wins 30 and player B wins 70.

Convert this constant-sum game to a zero-sum game.

Solution

The total pay-off for each cell is 100: the constant sum of the game is 100. Assume that each player puts half of this, 50 each, into a 'kitty' from which the pay-out is made.

Table 7.8 then becomes

	A	B
A	10, −10	0, 0
B	−20, 20	30, −30

Table 7.9

and this can be rewritten in the usual way for a zero–sum game:

	A	B
A	10	0
B	−20	30

Table 7.10

The pay-offs are for player 1, as is customary in a zero-sum game.

ACTIVITY 7.1

Convert the following pay-off matrix to zero-sum form, and find the value of the game.

	A	B
A	25, 75	65, 35
B	40, 60	60, 40

Table 7.11

Exercise 7.1

① The pay-off matrix for a zero-sum game between players 1 and 2 is as follows.

	A	B	C
A	2	0	1
B	−1	3	−4

Table 7.12

(i) Determine the pay-off matrix from player 2's point of view.

(ii) What is the most that player 2 can win?

(iii) What is the most that player 2 can lose?

② The pay-off matrix for a zero-sum game between players 1 and 2 is as follows.

	A	B	C
A	0	1	1
B	−1	5	2
C	4	−2	−3

Table 7.13

(i) What will be the outcome if both players play safe?

(ii) What will be the outcome if player 1 plays safe, and player 2 hears of player 1's intention?

(iii) What will be the outcome if (instead) player 2 plays safe, and player 1 hears of player 2's intention?

③ The pay-off matrix for a zero-sum game between players 1 and 2 is as follows.

	A	B	C
A	4	−1	2
B	0	−3	−2
C	−3	−2	6
D	−5	−4	3

Table 7.14

Show that this game has a stable solution, and give the value of the game.

④ For each of the following pay-off matrices for zero-sum games, determine the play-safe strategies for player 1 and player 2, or say if there is no stable solution.

(i)

	A	B
A	1	−1
B	2	0

Table 7.15

(ii)

	A	B	C	D
A	3	2	0	1
B	1	−2	−1	0

Table 7.16

(iii)

	A	B	C
A	−2	1	4
B	0	2	−1
C	3	−3	0

Table 7.17

(iv)

	A	B
A	1	3
B	2	0
C	4	−1

Table 7.18

⑤ For the pay-off matrices in question 4, identify any stable solutions and give the values of the games in these cases.

⑥ Show that if $a < b < c < d$, then a stable solution will exist for the zero-sum game pay-off matrix shown in Table 7.15.

	A	B
A	a	b
B	d	c

Table 7.19

2 Mixed strategies

Dominated strategies

Where a stable solution doesn't exist, it may be possible to devise a strategy for both players where neither player's choices are predictable. This is referred to as a **mixed strategy**, and is considered in the next section.

However, before adopting this strategy, there is one simplifying process that can often be applied. This involves the idea of **dominance**.

> **Note**
> ----------------
> Dominance may be **strict** or **weak**. Strict dominance requires that each element in the row (or column) is strictly greater (or less) than the corresponding element in the alternative row (or column).

	A	B	C
A	1	2	2
B	−2	1	2
C	3	−1	0

Table 7.20

Referring to the pay-off matrix in Table 7.20 (which, by convention, is from player 1's point of view), you see that player 2 would never choose option C, because option B would always be either better or of equal value (from player 2's point of view), whatever choice player 1 made. Column B is said to **dominate** column C, and the pay-off matrix can be reduced to that in Table 7.21.

	A	B
A	1	2
B	−2	1
C	3	−1

Table 7.21

Similarly, for the new matrix in Table 7.21, player 1 would never choose option B, as row A dominates row B, and you can reduce the matrix further to that in Table 7.22.

	A	B
A	1	2
C	3	−1

Table 7.22

As will be seen in Exercise 7.2, using the idea of dominance sometimes leads to a 1×1 matrix, which gives the value of the game directly.

Dominance is considered further in Section 7.3.

Optimal mixed strategies

In a mixed strategy the players select their options with a certain probability, so that their expected outcomes are optimised. Because the games are zero-sum, the expected outcome of one player is the expected outcome of the other, with the sign reversed.

A mixed strategy can be employed whenever the pay-off matrix is of order $2 \times n$ (where $n \geqslant 2$), after the matrix has been reduced due to any dominance.

You start by looking at the 2×2 case.

Example 7.2

Establish whether there is a stable solution for the zero-sum game given by the following pay-off matrix.

	A	**B**	**Row minima**
A	1	2	①
C	3	−1	−1
Column maxima	3	②	

Table 7.23

Solution

Max (row minima) = 1

Min (column maxima) = 2

As these values are not equal, there is not a stable solution.

> Compare the max (row minima) with the min (column maxima).

Note

In Table 7.23, you see that if player 1 plays safe, they choose option A, as this guarantees that the worst outcome is 1 (as opposed to a worst outcome of −1 if option C is chosen).

Similarly, player 2 will choose option B, as this guarantees a worst outcome of −2 (2 for player 1), compared with the worst outcome of −3 if option A is chosen.

This means that the actual outcome is 2 for player 1 and −2 for player 2. If player 1 knows that player 2 will always choose option B, then they are happy to choose option A (as 2 ⩾ −1). But if player 2 knows that player 1 will always choose option A, then player 2 will want to change their choice to option A (as 1 < 2, or −1 > −2 from player 2's point of view).

But if player 1 knows that player 2 will choose option A, then player 1 will now want to choose option C, and so on. It can be shown that a stable solution is never obtained.

The following problems occur when there is no stable solution.

■ Neither player can safely assume what the other player's choice will be.

■ Where the game is repeated, one player may take advantage of the other, if their choice becomes predictable.

It is possible to find a strategy for each player that avoids these problems by deliberately introducing an element of uncertainty, so that each player makes their choice with certain probabilities. This is the mixed strategy already mentioned.

For the 2 × 2 matrix in Table 7.23, suppose that player 1 adopts option A with probability p and option C with probability $1 - p$.

Player 1's expected pay-off depends on player 2's choice.

If player 2 chooses option A, then player 1's expected pay-off is

$1p + 3(1 - p) = 3 - 2p.$

If, instead, player 2 chooses option B, then player 1's expected pay-off is

$2p + (-1)(1 - p) = 3p - 1.$

These expected pay-offs are functions of p and can be shown graphically, as in Figure 7.1.

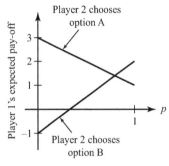

Figure 7.1

You see that player 1 can maximise their expected pay-off (obtaining 3) when player 2 chooses option A, by setting $p = 0$. However, they would risk receiving a pay-off of only -1 if player 2 chose option B. Similarly, if player 2 chooses option B, player 1's expected gain is maximised (obtaining 2) by setting $p = 1$, but they risk obtaining only 1 if player 2 chooses option A instead.

The safest strategy is to assume that player 2's choice always results in the worst possible outcome for player 1, so that whatever you set p equal to, player 1's expected pay-off is given by the lower of the two lines.

This means that you need to look for the value of p such that the lower of the two lines is as high as possible, and this occurs where the two lines intersect.

Example 7.3

Find the required value of p for the zero-sum game shown by the pay-off matrix in Table 7.23.

> **Note**
> ----------
> It does not matter whether the game is repeated: the expected pay-off for each game is 1.4, regardless of the number of games played.

Solution

$$3 - 2p = 3p - 1 \Rightarrow 4 = 5p \Rightarrow p = 0.8$$

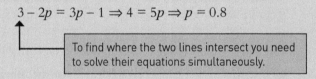

To find where the two lines intersect you need to solve their equations simultaneously.

With this value for p, player 1 does not mind what player 2 does: if the game is repeated, player 2 might choose the same option all the time, or they might themselves choose their option according to a probability rule.

Player 1's expected pay-off is $3 - 2p$ (or $3p - 1$) = 1.4.

Example 7.4

Find the probability rule that player 2 should adopt, and their expected pay-off.

Solution

Suppose that player 2 adopts option A with probability q and option B with probability $1 - q$.

If player 1 chooses option A, then player 2's expected pay-off is

$$-\{1q + 2\,(1 - q)\} = q - 2.$$

> The minus sign is needed because the values in the pay-off matrix show player 2's losses, not gains.

If, instead, player 1 chooses option C, then player 2's expected pay-off is

$$-\{3q + (-1)\,(1 - q)\} = 1 - 4q.$$

It is not essential to draw the graphs. However, they are shown in Figure 7.2.

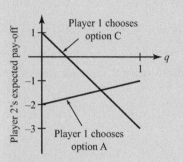

Figure 7.2

You then find the required value of q from $q - 2 = 1 - 4q \Rightarrow 5q = 3 \Rightarrow q = 0.6.$

Player 2's expected pay-off is $q - 2$ (or $1 - 4q$) $= -1.4.$

> As this is a zero-sum game, it is no surprise that player 2's expected pay-off is just player 1's expected pay-off with the sign reversed.

Note

Where a mixed strategy is employed, the value of the game is defined to be player 1's expected pay-off; in this case, 1.4.

$2 \times n$ and $n \times 2$ matrices (where $n > 2$)

This method can be extended to $2 \times n$ matrices: i.e. where player 2 has n choices.

In the case of $n \times 2$ matrices, where player 1 has n choices, the method cannot be used to find player 1's probability rule, as this would require more than one probability variable (instead of the single p).

However, the value of the game can be established by swapping the roles of player 1 and player 2, to obtain a $2 \times n$ matrix, and then find player 1's probability rule.

Example 7.5

Find expressions, in terms of p, for player 1's expected pay-off, for each of player 2's three options, for the pay-off matrix in Table 7.24.

Draw the graphs of the corresponding straight lines on a single diagram.

	A	B	C
A	0	−1	2
B	2	3	−2

Table 7.24

Discussion point

→ In some situations, the lowest of the lines is at the highest level when either $p = 0$ or 1. What can you say about such cases? Consider the pay-off matrix $\begin{pmatrix} 4 & 1 \\ 3 & 2 \end{pmatrix}$.

Solution

The first step is always to check whether the matrix can be reduced, due to any dominance of either rows or columns. Here there is no dominance.

As before, suppose that player 1 adopts option A with probability p and option B with probability $1 - p$.

If player 2 chooses option A, then player 1's expected pay-off is

$0p + 2(1 - p) = 2 - 2p.$

If player 2 chooses option B, then the expected pay-off is

$(-1)p + 3(1 - p) = 3 - 4p.$

If player 2 chooses option C, the expected pay-off is

$2p + (-2)(1 - p) = 4p - 2.$

The graphs of the three straight lines are shown in Figure 7.3.

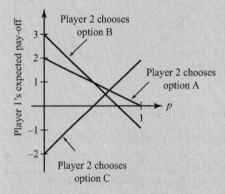

Figure 7.3

Once again, you need to look for the value of p such that the lower of the lines is as high as possible. This occurs where the lines for options B and C intersect. Note that it was necessary to draw the graphs in order to see this.

Hence $3 - 4p = 4p - 2$, so that $5 = 8p$ and $p = \frac{5}{8}$.

With this value of p, the expected pay-off for player 1 is $3 - 4p$ (or $4p - 2$) $= \frac{1}{2}$ (this is the value of the game).

The following example shows a device that can be used to find player 2's probability rule, where player 2 has three options.

Example 7.6

Find player 2's probability rule for the pay-off matrix (Table 7.24) of Example 7.5.

Solution

Suppose that player 2 adopts options A, B and C with probabilities q, r and $1 - q - r$, respectively.

If player 1 chooses option A, then player 2's expected pay-off is

The minus sign is needed because the values in the pay-off matrix show player 2's losses, not gains.

$-\{0q + (-1)\,r + 2\,(1 - q - r)\} = -2 + 2q + 3r.$

If, instead, player 1 chooses option B, then player 2's expected pay-off is

$-\{2q + 3r + (-2)\,(1 - q - r)\} = 2 - 4q - 5r.$

As usual, the probability rule is chosen in such a way that the expected pay-offs are the same, whichever option player 1 chooses, and both are equal to the value of the game from player 2's point of view. (This corresponds to the intersection of the lines in the simple 2×2 case.)

So $-2 + 2q + 3r = 2 - 4q - 5r = -\frac{1}{2}$ (as the value of the game from player 1's point of view is $\frac{1}{2}$, from Example 7.4).

These equations can be written as

$-4 + 4q + 6r = -1$ and $4 - 8q - 10r = -1$
or $4q + 6r = 3$ ① and $8q + 10r = 5$ ②.

$2 \times$ ① $-$ ② gives $2r = 1$, so that $r = 0.5$, and $q = 0$

Remember to always state the probabilities at the end of your calculation.

Thus the probability rule for player 2 is to never choose option A and choose options B and C with equal probability (0.5).

Exercise 7.2

① Consider the following pay-off matrix.

	A	B	C	D
A	2	1	3	3
B	0	0	−1	0
C	1	1	0	0

Table 7.25

(i) Which option(s) should player 1 never choose?

(ii) Which option(s) should player 2 never choose?

② For each of the pay-off matrices in question 4 of Exercise 7.1, identify any dominated strategies, and give the reduced matrix in these cases.

③ Show that there is no stable solution for the following pay-off matrix.

	A	B
A	3	2
B	1	4

Table 7.26

Use a mixed strategy to find the value of the game.

④ (i) Consider the game with the following pay-off matrix.

	A	B
A	1	2
B	3	0
C	−1	0
D	−1	4

Table 7.27

Are there any choices that either player should never make?

(ii) Find the value of the game.

⑤ Use a mixed strategy to find the values for player 1 of the games having the following pay-off matrices.

(i)

	A	B
A	1	4
B	5	3

Table 7.28

(ii)

	A	B	C
A	−1	2	5
B	9	7	3

Table 7.29

(iii)

	A	B
A	−1	5
B	2	4
C	7	3

Table 7.30

6 (i) Show that there is a stable solution for the following pay-off matrix, without reducing the matrix. Give the value of the game.

	A	B	C	D	E
A	1	0	−1	2	3
B	1	2	0	1	2
C	2	1	−1	4	3

Table 7.31

(ii) Reduce the pay-off matrix, as far as possible.

7 Reduce the following pay-off matrix as much as possible, and show that there is no stable solution to the game.

	A	B	C	D
A	2	−1	0	4
B	0	6	2	2
C	4	0	1	1
D	1	−3	−1	2
E	−2	3	1	−3

Table 7.32

8 For what values of x can the following pay-off matrix be reduced?

	A	B	C
A	2	x	1
B	1	4	0

Table 7.33

9 Determine optimal strategies for both players for the game with the following pay-off matrix, and draw a graph to illustrate the situation from player 1's point of view.

	A	B
A	3	−1
B	2	4

Table 7.34

10 Determine the value of the game with the following pay-off matrix.

	A	B	C
A	5	2	−3
B	−1	4	6

Table 7.35

11 Determine optimal strategies for both players for the game with the following pay-off matrix, and establish the value of the game.

	A	B	C
A	4	3	2
B	−2	1	4
C	2	1	5

Table 7.36

3 Nash equilibrium

You shall now consider the more general case of non-zero sum games.

An important idea in game theory is that of a **Nash equilibrium**. It is assumed that each player has full knowledge of the pay-off matrix, and a particular combination of choices is said to constitute a Nash equilibrium if neither player would want to change their choice, assuming that the other player maintained their choice.

Example 7.7

The potential winnings of the two players are as shown in the following pay-off matrix:

	A	B
A	8, 7	1, 9
B	10, 2	4, 3

Table 7.37

This means that if player 1 chooses option A and player 2 chooses option B, for example, then player 1 wins 1 and player 2 wins 9.

Find any Nash equilibrium solutions.

Note

Strict Nash equilibrium occurs when a player will not want to change their choice because the alternative is inferior for them. However, a situation might arise where a player is indifferent between the various options (i.e. the gains are the same), but where a change of choice would be detrimental to the other player. This is referred to as non-strict Nash equilibrium.

Solution

If player 1 chooses A and player 2 chooses A ('choices A, A'), then player 1 would prefer to change to option B (assuming that player 2 still chooses A), as $10 > 8$. Also, player 2 would prefer to change to option B (assuming that player 1 still chooses A), as $9 > 7$.

For A, B, player 1 would prefer to change (as $4 > 1$), although player 2 would not (as $7 < 9$).

For B, A, player 1 would not want to change (as $8 < 10$), although player 2 would (as $3 > 2$).

For B, B, player 1 would not want to change (as $4 > 1$), and neither would player 2 (as $3 > 2$).

Thus, of the four cases, only B, B constitutes a Nash equilibrium.

Note

An alternative way of finding Nash equilibrium positions (which can be shown to be equivalent) is as follows.

Suppose that player 1 chooses A. Then player 2 would prefer to choose B, so circle the 9 in cell A, B. If, instead, player 1 chooses B, then player 2 would prefer to choose B, so circle the 3 in cell B, B. If player 2 chooses A, then player 1 would prefer to choose B, so circle the 10 in cell B, A. If, instead, player 2 chooses B, then player 1 would prefer to choose B, so circle the 4 in cell B, B.

This gives the following table.

	A	**B**
A	8, 7	1, ⑨
B	⑩, 2	④, ③

Table 7.38

A Nash equilibrium then occurs wherever both of the numbers in a cell are circled.

Note, however, that a Nash equilibrium is not necessarily the best solution for the two players. In the previous example, A, A is better than B, B for both players.

This situation is known as the **Prisoners' dilemma**. One version concerns two accomplices in a crime. Option A is to confess their part in the crime, whilst option B is to inform on their accomplice. The winnings in this case are the reductions in their sentences.

Strict and weak dominance

As has been seen on page 128, a row (column) is said to strictly dominate another row (column) if each of its elements is strictly greater than the corresponding element in the other row.

Where at least one of the elements is the same as the corresponding element in the other row (and there is dominance), this is referred to as weak dominance.

In the case of weak dominance, it is possible to lose a non-strict Nash equilibrium.

For example, consider the following pay-off matrix for a zero-sum game.

	A	**B**
A	3	4
B	3	5

Table 7.39

Row B weakly dominates row A. Player 1 can safely ignore option A, and the pay-off matrix then reduces to

	A	B
B	3	5

Table 7.40

Now column A dominates column B (as player 2 prefers option A), and a stable solution is therefore B, A.

However, A, A can be seen to be a (non-strict) Nash equilibrium, and this has been lost in the reduction process.

Also, in the case of weak dominance, the order of the reductions may affect the outcome.

In Table 7.39, column A dominates column B, and so the pay-off matrix can be reduced to

	A
A	3
B	3

Table 7.41

In this situation, neither row is said to dominate the other, and both A, A and B, A are stable solutions.

Exercise 7.3

① Find any (strict) Nash equilibrium positions for the following pay-off matrices

(i)

	A	B
A	9, 8	5, 7
B	7, 6	8, 10

Table 7.42

(ii)

	A	B
A	5, 7	3, 7
B	6, 4	1, 6

Table 7.43

(iii)

	A	B
A	10, 8	7, 7
B	6, 7	8, 5

Table 7.44

(iv)

	A	B
A	6, 9	4, 6
B	8, 6	2, 7

Table 7.45

② Show that a stable solution for a zero-sum game will be a Nash equilibrium, using the following pay-off matrix as an illustration

	A	B
A	9	1
B	10	2

Table 7.46

③ For the following zero-sum pay-off matrix, establish whether the order of any possible reductions affects the outcome. Also establish whether any Nash equilibrium is lost.

	A	B
A	2	3
B	1	2
C	2	4

Table 7.47

4 Converting games to linear programming problems

So far, you have established methods for dealing with the following situations.

■ 2 × 2 pay-off matrices

■ cases where there is dominance, so that the pay-off matrix can be reduced

■ 2 × n or n × 2 pay-off matrices.

For other orders of matrices, you can use linear programming methods.

In order to illustrate the method, consider the pay-off matrix in Table 7.48. This uses the same numbers as in Table 7.24 (swapping the roles of the two players in Table 7.24), and the value of the game can be found to be $\frac{1}{2}$.

Note

This is a new example, where player 1 now has three options.

	A	B
A	0	2
B	−1	3
C	2	−2

Table 7.48

Suppose that player 1 chooses the three options with probabilities p_1, p_2 and p_3.

Let v be the value of the game.

In order for the simplex method to work, you require the control variables to be non-negative. The probabilities are non-negative, but you also want $v \geqslant 0$. Suppose that the pay-off matrix is of the general form shown in Table 7.49, and that any dominance has been removed.

	A	B
A	λ_1	μ_1
B	λ_2	μ_2
C	λ_3	μ_3

Table 7.49

Then, as before, v will be found from the intersection of $\lambda_1 p_1 + \lambda_2 p_2 + \lambda_3 p_3$ and $\mu_1 p_1 + \mu_2 p_2 + \mu_3 p_3$,

so that $v = \lambda_1 p_1 + \lambda_2 p_2 + \lambda_3 p_3 = \mu_1 p_1 + \mu_2 p_2 + \mu_3 p_3$ (*)

In order for the simplex method to work, you need to have positive values for λ_i and μ_i. The reason for this will be explained shortly.

As it is only the relative values of the elements that are important, you can add any amount to all of the elements, provided you subtract it from the expected pay-off. For the present example, you can add 3 to each element, to produce the matrix in Table 7.50.

	A	B
A	3	5
B	2	6
C	5	1

Table 7.50

Using this new matrix, the expected pay-offs for player 1 are $3p_1 + 2p_2 + 5p_3$ or $5p_1 + 6p_2 + 1p_3$, depending on whether player 2 chooses option A or option B.

As before, assume the worst case scenario: namely that, whatever probabilities are adopted, player 2's choice will always result in v being the lower of $3p_1 + 2p_2 + 5p_3$ and $5p_1 + 6p_2 + 1p_3$. Given this constraint, aim to maximise v.

Also, $p_1 + p_2 + p_3 = 1$, but for the purpose of linear programming an inequality is required. However, it is possible to write $p_1 + p_2 + p_3 \leqslant 1$, for the following reason.

Suppose that v has been maximised with $p_1 + p_2 + p_3 < 1$.

Then it would be possible to increase one of the probabilities, and thereby increase both $3p_1 + 2p_2 + 5p_3$ and $5p_1 + 6p_2 + 1p_3$, so that the lower of these expressions would be increased, giving a higher value for v, which contradicts the assumption that v has been maximised. So $p_1 + p_2 + p_3 < 1$ isn't possible, and therefore $p_1 + p_2 + p_3 = 1$.

Note that, in the above argument, when one of the probabilities is increased, the expressions for v (in general, $\lambda_1 p_1 + \lambda_2 p_2 + \lambda_3 p_3$ and $\mu_1 p_1 + \mu_2 p_2 + \mu_3 p_3$) can only be guaranteed to increase if λ_1 and p_1 are all positive.

The linear programming problem can therefore be expressed as

maximise $p = v - 3$ subject to

$v \leqslant 3p_1 + 2p_2 + 5p_3$
$v \leqslant 5p_1 + 6p_2 + 1p_3$
$p_1 + p_2 + p_3 \leqslant 1$
$p_1, p_2, p_3 \geqslant 0$.

This can be solved using the simplex method.

Example 7.8

Use the simplex method to find the value of the game with the following pay-off matrix.

	A	B
A	1	2
B	3	−1

Table 7.51

> You have already found this value to be 1.4 in Example 7.2.

Solution

Add 2 to each element, so that all elements are positive, to give the following matrix.

	A	B
A	3	4
B	5	1

Table 7.52

Maximise $P = v - 2$, subject to the constraints

$v \leqslant 3p_1 + 5p_2, v \leqslant 4p_1 + p_2, p_1 + p_2 \leqslant 1$ $(p_1, p_2 \geqslant 0, v > 0)$.

Introducing slack variables, the simplex equations are

$P - v = -2$ $\qquad$ $v - 4p_1 - p_2 + s_2 = 0$

$v - 3p_1 - 5p_2 + s_1 = 0$ $\qquad$ $p_1 + p_2 + s_3 = 1$

The simplex tableau is shown in Table 7.53.

P	v	p_1	p_2	s_1	s_2	s_3	Value	Equation
1	-1	0	0	0	0	0	-2	①
0	①	-3	-5	1	0	0	0	②
0	1	-4	-1	0	1	0	0	③
0	0	1	1	0	0	1	1	④
1	0	-3	-5	1	0	0	-2	⑤ = ① + ⑥
0	1	-3	-5	1	0	0	0	⑥ = ②
0	0	-1	④	-1	1	0	0	⑦ = ③ − ⑥
0	0	1	1	0	0	1	1	⑧ = ④
1	0	$-\dfrac{17}{4}$	0	$-\dfrac{1}{4}$	$\dfrac{5}{4}$	0	-2	⑨ = ⑤ + 5 × ⑪
0	1	$-\dfrac{17}{4}$	0	$-\dfrac{1}{4}$	$\dfrac{5}{4}$	0	0	⑩ = ⑥ + 5 × ⑪
0	0	$-\dfrac{1}{4}$	1	$-\dfrac{1}{4}$	$\dfrac{1}{4}$	0	0	⑪ = ⑦ ÷ 4
0	0	$\boxed{\dfrac{5}{4}}$	0	$\dfrac{1}{4}$	$-\dfrac{1}{4}$	1	1	⑫ = ⑧ − ⑪
1	0	0	0	$\dfrac{3}{5}$	$\dfrac{2}{5}$	0	$\dfrac{7}{5}$	⑬ = ⑨ + $\dfrac{17}{4}$ × ⑯
0	1	0	0	$\dfrac{3}{5}$	$\dfrac{2}{5}$	0	$\dfrac{17}{5}$	⑭ = ⑩ + $\dfrac{17}{4}$ × ⑯
0	0	0	1	$-\dfrac{1}{5}$	$\dfrac{1}{5}$	$\dfrac{1}{5}$	$\dfrac{1}{5}$	⑮ = ⑪ + $\dfrac{1}{4}$ × ⑯
0	0	1	0	$\dfrac{1}{5}$	$-\dfrac{1}{5}$	$\dfrac{4}{5}$	$\dfrac{4}{5}$	⑯ = ⑫ × $\dfrac{4}{5}$

Table 7.53

The maximised value of P is $\dfrac{7}{5} = 1.4$, as before.

Also, as a check, $p_1 + p_2 = 1$, where $p_1 = \dfrac{4}{5}$ and $p_2 = \dfrac{1}{5}$, also as before.

Exercise 7.4

① Formulate the linear programming problem that can be used to find the value of the game with the following pay-off matrix.

	A	B
A	1	3
B	2	0
C	4	−1

Table 7.54

② Use the simplex method to find the value of the game in question 3 of Exercise 7.2, with the following pay-off matrix.

	A	B
A	3	2
B	1	4

Table 7.55

③ Use the simplex method to find the values of the games in question 5 of Exercise 7.2.

④ Formulate the linear programming problem that can be used to find the value of the game with the following pay-off matrix. Write down the initial simplex equations.

	A	B
A	−2	−1
B	−3	0
C	1	−2

Table 7.56

⑤ Establish the initial simplex tableau that can be used to find the value of the game with the following pay-off matrix.

	A	B
A	4	−3
B	−1	2

Table 7.57

⑥ For what value(s) of x is there a stable solution for the following pay-off matrix?

	A	B
A	2	x
B	0	3
C	1	2

Table 7.58

⑦ A mathematical board game for two players is being trialled. When player 1 lands on a particular square, the amount he or she wins is the highest common factor of two numbers. These numbers are chosen from 3 specified numbers, with each player choosing one of the numbers (they can both choose the same number). Player 2 is trying to minimise the amount that player 1 wins. Janet and John are playing this game, and Janet lands on a square with the numbers 24, 30 and 45.

(i) Which numbers should Janet and John choose if they adopt play safe strategies, and how much will Janet win in that case?

(ii) If (unknown to Janet) John has forgotten how to work out a highest common factor, and chooses his number at random, what amount can Janet expect to win, on average?

LEARNING OUTCOMES

Now you have finished this chapter, you should be able to

➤ understand, interpret and construct pay-off matrices for zero-sum games

➤ find play-safe strategies and the value of the game

➤ prove the existence or non-existence of a stable solution

➤ identify and make use of dominated strategies

➤ find optimal mixed strategies for a game, including use of graphical methods

➤ understand the concept of a Nash equilibrium

➤ convert a constant-sum game to a zero-sum game

➤ convert higher-order games to linear programming problems, and solve them.

KEY POINTS

1 In a zero-sum game between two players, if player 1 wins a certain amount, then player 2 loses that amount.
2 The pay-off matrix shows the possible gains from player 1's point of view.
3 A play-safe strategy is where a player chooses the option with the smallest downside.
4 A stable solution exists when it is in both players' interests to adopt the play-safe strategy. This occurs when the maximum of the row minima equals the minimum of the column maxima.
5 A pay-off matrix can be reduced if one strategy dominates another.
6 In a mixed strategy, the players select their options with a certain probability in such a way that their expected outcomes are optimised.
7 The value of a game is player 1's pay-off, where there is a stable solution; or their expected pay-off when a mixed strategy is employed.
8 A mixed strategy can be employed whenever the (reduced) pay-off matrix is of order $2 \times n$ or $n \times 2$.
9 The players' choices constitute a Nash equilibrium if neither player would want to change their choice, assuming that the other player maintained their choice.
10 A Nash equilibrium is not necessarily the best solution for the two players.
11 It is possible to convert from a constant-sum game to a zero-sum game.
12 In more complicated cases, the Simplex method may be employed.

Answers

Chapter 1

Opening activity Page 1

8	1	6
3	5	7
4	9	2

Activity 1.1 Page 2

For example,

(i) Is there a route from A to D?

(ii) Find a route from A to D.

(iii) How many routes are there from A to D?

(iv) Which is the shortest route from A to D?

Activity 1.2 Page 3

6

Exercise 1.1 Page 4

1 (i) Existence problem

  (ii) For example, Sehrish is researching the cheapest way to get from Newcastle to London in less than 2 hours.

2 There are 10 different possible end digits (pigeonholes) and 11 digits to be placed, so at least one of the digits must occur more than once.

3 For example,

  (i) Do three of the lines form an equilateral triangle?

  (ii) Find a route that passes through all 7 points.

  (iii) How many lines are there in Figure 1.2?

  (iv) What is the minimum total length of lines that joins all of the 7 points?

4 There are 4 odd numbers and 4 even numbers. A set of 5 integers must include at least one odd number and at least one even number. Odd + even = odd.

5 The mean value of the integers 1 to 10 is 5.5. The mean sum of three numbers is 16.5 so the maximum sum must be at least 16.5. Since they are all integers this maximum sum must be at least 17.

Activity 1.3 Page 5

(i) (a) 1, 2, 3, 4, 6, 12

  (b) {5, 7, 8, 9, 10, 11}

  (c) 6

(ii) Factors of 12

Activity 1.4 Page 6

52

Exercise 1.2 Page 7

1 (i) (b), (d)

  (ii) (a), (e)

  (iii) (c), (f)

2 20

3 No, because $\mathbb{Z} \supset \mathbb{Q}$ and also irrational numbers are not included in the partition.

4 19

5 38

Activity 1.5 Page 7

ABCD, ABDC, ACBD, ACDB, ADBC, ADCB

BACD, BADC, BCAD, BCDA, BDAC, BDCA

CABD, CADB, CBAD, CBDA, CDAB, CDBA

DABC, DACB, DBAC, DBCA, DCAB, DCBA

Activity 1.6 Page 8

10 000

Exercise 1.3 Page 10

1 362 880

2 336

3 60

4 126

5 (i) 288

  (ii) 144

6 16

7 48

8 96

9 (i) 181 440

  (ii) 141 120

10 (i) 3 268 760

  (ii) 1 019 304

Activity 1.7 Page 11

$5 + 4 + 5 - 1 - 2 - 2 + 1 = 10$

Exercise 1.4 Page 12

1 26

2 192

3 44

4 120

5 13
6 265
7 5292
8 44
9 12
10 (i) 34 560
   (ii) 777 600

Chapter 2

Opening activity Page 14

Yes

Discussion point Page 16

The smallest number of edges occurs when the graph is a tree, when the number of edges will be $n - 1$.

Activity 2.1 Page 17

Every edge contributes two to the sum of the degrees of the vertices, so that this sum must be even. As the sum of the degrees of the even vertices is even, the sum of the degrees of the odd vertices must also be even, in order to give an even sum overall. So the number of odd vertices must be even.

Exercise 2.1 Page 17

1 (i)   ABCA, ABCDEA, ACEA, ACDEA, ABCEA
   (ii) ABCEDCA is not a cycle since vertex C is repeated.
   (iii) A (closed) trail. It is a walk too, but since no edge is repeated it is best described as a trail.
2 Possible examples are:

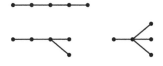

3 (i)

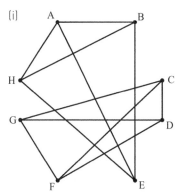

   (ii) $^{8}C_{2} = 28$

4

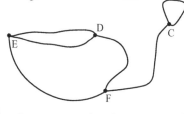

5

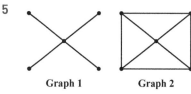

   Graph 1   Graph 2

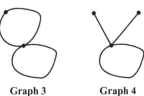

   Graph 3   Graph 4

6 (i) $14 = 2 \times 7$, because each of the 7 edges has 2 ends.
   (ii) Two of the following:

	Number of degree 2	Number of degree 3	Number of degree 4
(a)	3	0	2
(b)	2	2	1
(c)	1	4	0

   (iii) Possible examples for each of the possibilities:
   (a)   (b)   (c)

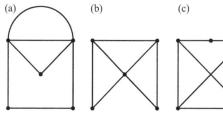

7 (i) $6 - 21; 4 - 14; 2 - 7$

   (ii) $d = 2, 4, 5, 6$ or 7
   (iii) The number of edges $= \frac{1}{2} \times$ no. of nodes $\times$ degree of the nodes. Since the number of nodes is odd, the degree must be even to have an integer value for the number of edges. When there are 8 vertices, the number of edges will be an integer, whether d is odd or even.

Discussion point Page 21

mn

Discussion point Page 21

Because the vertices on one side are not connected to each other.

Activity 2.2 Page 21

2

Exercise 2.2 Page 21

1   Graph 1 – (ii)
     Graph 2 – (i)
     Graph 3 – (iii)
     Graph 4 – (ii)

2

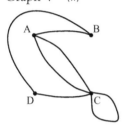

3

	A	B	C	D	E
A	2	0	0	0	0
B	0	0	1	1	1
C	0	1	0	0	0
D	0	1	0	0	1
E	0	1	0	1	0

4   (i)

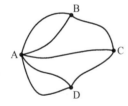

(ii) The four vertices of the graph are all odd, so that the graph is neither Eulerian nor semi-Eulerian.

5   (i)

(ii)

(iii) Three connection points and three internal points

6

	A	B	C	D	E	F	G
A	–	7	–	–	–	11	20
B	7	–	10	8	–	–	15
C	–	10	–	7	–	–	–
D	–	8	7	–	6	–	9
E	–	–	–	6	–	8	7
F	11	–	–	–	8	–	12
G	20	15	–	9	7	12	–

7

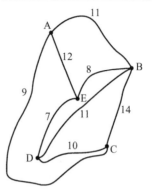

8   (i) It is Eulerian when it has an odd number of vertices, as that makes each vertex even. It can only be semi-Eulerian when there are two odd vertices.

(ii) It is Eulerian when each set has an even number of vertices; it is semi-Eulerian when one set has two vertices and the other has an odd number of vertices, or if each set has one vertex.

9   For example, all the possible edges are drawn in. To draw an edge you choose two vertices to join. So the number of edges is $^nC_2 = \frac{1}{2}n\,(n-1)$
Or: Each of the *n* vertices has $n-1$ edges leading from it, giving a total of $n(n-1)$ edges. However, each of the edges is being counted twice (once for each of the vertices at its ends), and so you divide by 2 to remove the duplication, giving $\frac{1}{2}n(n-1)$ edges.

10   (i) AB; ACB; ADB; ACDB; ADCB
     (ii) 1 direct, AB
          3 through one other point
          $^3P_2 = 6$ through 2 other points
          $3! = 6$ through 3 other points
          $1 + 3 + 6 + 6 = 16$

Exercise 2.3 Page 23

1   Label the first matrix A, B, C, D. Label the second matrix C, A, D, B and it can clearly be seen that they are isomorphic.

2   (2) and (6); (1), (3), (4) and (5)

3 (i) For example A ↔ 1; D ↔ 2;
    E ↔ 3; B ↔ 5; C ↔ 6; F ↔ 4

(ii)
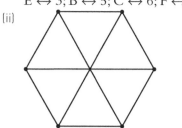

4 B is a 4-node and connects to C which is a
2-node in the first graph. The only 2-node in
the second graph, U, does not connect with the
only 4-node, which is Q and so they cannot be
isomorphic.

5 P – B; Q – E; R – A; S – C; T – D

Activity 2.3 Page 25
Penzance

Activity 2.4 Page 26
ADEBCFA

Exercise 2.4 Page 26
1 Graphs 1, 3 and 4
2 For example:

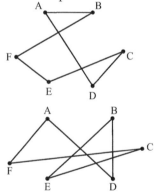

3 24
4 Graph 1: deg(A) + deg(D) = 2 + 2 < 6, so cannot
conclude that graph is Hamiltonian from Ore's
theorem; but graph is in fact Hamiltonian
(e.g. CAEDBFC).
Graph 2: deg(A) + deg(F) = 3 + 3 < 7, so cannot
conclude that graph is Hamiltonian from Ore's
theorem; graph is not Hamiltonian.
Graph 3: degree of each vertex is ≥ 3, so deg
v + deg w ≥ 6 and Ore's theorem ⇒ graph is
Hamiltonian.
Graph 4: deg(A) + deg(D) = 2 + 2 < 5, so cannot
conclude that graph is Hamiltonian from Ore's
theorem; but graph is in fact Hamiltonian
(e.g. DECABD).

Activity 2.5 Page 28
$V + R = 11 + 10 = 21$
$E + 2 = 19 + 2 = 21$

Discussion point Page 29
Yes, because K_4 counts as a subdivision of itself.

Activity 2.6 Page 31

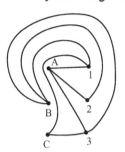

Exercise 2.5 Page 33
1 Graph 1 is planar as it does not contain a
subdivision K_5.
Graph 2 is planar as it does not have a subgraph
that is K_5 or $K_{3,3}$.
Graph 3 is not planar as K_5 is a subgraph of it
(remove B or D).
Graph 4 is not planar as it has $K_{3,3}$ as a subgraph (lose
C and 4).

2 Graph 1: $V = 6, R = 6, E = 10$
$V + R = 6 + 6 = 12$
$E + 2 = 10 + 2 = 12$
Graph 2: $V = 6, R = 7, E = 11$
$V + R = 6 + 7 = 13$
$E + 2 = 11 + 2 = 13$

3 K_5 has 10 edges so a graph with 8 edges does not
have a subgraph that is K_5. $K_{3,3}$ has 9 edges so a
graph with 8 edges does not contain a subgraph
that is $K_{3,3}$. Hence result.

4 (i) $K_{4,2}$ does not have a subgraph that is $K_{3,3}$, as
there are only two vertices in the second set
of vertices.

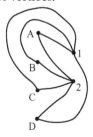

(ii) $K_{n,2}$ does not have a subgraph that is $K_{3,3}$,
as there are only two vertices in the second
set of vertices.

5

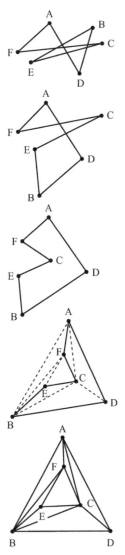

6 (i) No (ii) No (iii) Yes
(iv) No (v) Yes

7 Yes (The graph doesn't contain a subgraph that is a subdivision of K_5.)

8 (i) (a) a–c1; b–c2; c–c3; d–c1; e–c4
(b) a–c1; b–c2; c–c3; d–c2; e–c1
(ii) (a) The vertex would have to be coloured differently to itself!
(b) One edge is enough to force a different colour. A second adds nothing.
(iii) (a)

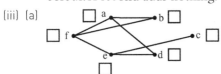

(b) a–c1; b–c2; c–c1; d–c2; e–c3; f–c4
(c) colour = hour;
a & c then b & d then e then f
(d) e.g. a & c then b & e then d & f

9 Draw a complete graph on six vertices. Imagine colouring the edge connecting two vertices green if the people represented by the vertices know each other, and red if they do not. Choose a vertex. It has 5 edges incident upon it, so there must be three of the same colour, c1 say. Now examine the three edges joining vertices at the other ends of those three. If they are all coloured c2, then there is a c2 triangle. If not, then at least one is coloured c1, and that means that there is a c1 triangle which includes the original vertex.

10 Every person must map to a number. If there are n people then the image set is $\{0, 1, 2, ..., n-1\}$. But 0 and $n-1$ cannot both be images since it is not possible simultaneously to have had somebody shake hands with everybody, and somebody else shake hands with nobody. Thus there are $n-1$ possible images for n subjects, so the mapping is many–one.

Chapter 3

Opening activity Page 36

2098

Activity 3.2 Page 37

Step	N	A	B	Comments	Passes
1	2				
2		1		$A = \frac{1}{2} \times 2$	
3			1.5	$B = \frac{1}{2}(1 + (2 \div 1))$	First pass
4				$(A - B)^2 = (1 - 1.5)^2 = 0.25$	
5		1.5		(new) A = (old) B	
3			1.41667	$B = \frac{1}{2}(1.5 + (2 \div 1.5))$	Second pass
4				$(1.5 - 1.41667)^2 = 0.00694$	
5		1.41667			
3			1.41422	$(1.41667 - 1.41422)^2 \leq 0.001$	Third pass
6				DISPLAY 1.41422	

Activity 3.3 Page 39

Russian algorithm		Euclidean algorithm	
13	37		
~~26~~	~~18~~	x	y
52	9	6	15
~~104~~	~~4~~	6	9
~~208~~	~~2~~	6	3
<u>416</u>	1	3	3
481		Output value 3	

Exercise 3.1 Page 41

1

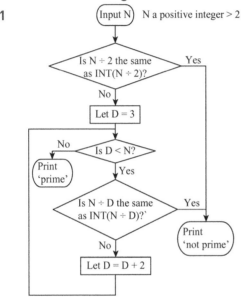

(Input N) N a positive integer > 2

Is N ÷ 2 the same as INT(N ÷ 2)? — Yes

No

Let D = 3

Is D < N? — No → Print 'prime'

Yes

Is N ÷ D the same as INT(N ÷ D)? — Yes → Print 'not prime'

No

Let D = D + 2

2 (i) Sixty-two 8 eight 5 five 4 four 4

   (ii) $\{1, 2, 6, 10\} \rightarrow 3, 5, 4$
   $\{4, 5, 9\} \rightarrow 4$
   $\{3, 7, 8\} \rightarrow 5, 4$
   $\{11, 12\} \rightarrow 6, 3, 5, 4$

3 (i) $6 \rightarrow 3 \rightarrow 10 \rightarrow 5 \rightarrow 16 \rightarrow 8 \rightarrow 4 \rightarrow 2 \rightarrow 1$
   $\rightarrow 4 \rightarrow 2 \rightarrow 1 \rightarrow$ ...(can stop at second '4')

   (ii) $256 \rightarrow 128 \rightarrow 64 \rightarrow 32 \rightarrow 16 \rightarrow 8 \rightarrow 4 \leftarrow 2$
   $\rightarrow 1 \rightarrow$...(As above or can note repetition from '16'.)

   (iii) e.g. Step 25 If n is 1 then stop. (Any step number between 21 and 29, or indicated in some other way.)

   (iv) Need to know that all chosen numbers lead to 1.

4 (i)

n	1	2	3	4	5	6	7
p	1	0.962	0.888	0.785	0.664	0.537	0.413

   (ii) Need to collect 7 cards for the probability of repetition on the list to exceed 0.5.

(iii)

Step 1	Set $n = 1$
Step 2	Set $p = 1$
Step 3	Set $n = n + 1$
Step 4	Set $p = p \times \frac{(366 - n)}{365}$
Step 5	If $p < 0.5$ then stop
Step 6	Go to Step 3

(iv) Because they do not have the same frequency of occurrence (probability) as other birthdays.

5 (i) e.g.

   Step 1 Input N, the number of symbols in the Roman numeral

   Step 2 Let A = 1, R = 1 and T = 0

   Step 3 Read the symbol in position A of the Roman numeral, call this S

   Step 4 Read the cell in row R and the column labelled S. Increase T by the first number in the cell and change R to the second number in the cell.

   Step 5 Increase A by 1

   Step 6 If A > N, display T and stop

   Step 7 Otherwise go to Step 3

   (ii) Roman numeral must be less than 5000

   (iii) e.g.

   Step 1 Input N {N a positive integer < 5000}

   Step 2 U = N − 10 × INT(N ÷ 10)

   Step 3 If U < 5 let I = U
   If U > 5 let V = 1 and I = U − 5

   Step 4 N = (N − U) ÷ 10

   Step 5 T = N − 10 × INT(N ÷ 10)

   Step 6 If T < 5 let X = T
   If T > 5 let L = 1 and X = T − 5

   Step 7 N = (N − T) ÷ 10

   Step 8 H = N − 10 × INT(N ÷ 10)

   Step 9 If H < 5 let C = H
   If H > 5 let D = 1 and C = H − 5

   Step 10 N = (N − H) ÷ 10

   Step 11 M = N − 10 × INT(N ÷ 10)

   Step 12 Display M copies of the symbol 'M'

   Step 13 If C = 4 and D = 0 display 'CD'
   If C = 4 and D = 1 display 'CM'
   Otherwise display D copies of 'D' followed by C copies of 'C'

   Step 14 If X = 4 and L = 0 display 'XL'
   If X = 4 and L = 1 display 'XC'
   Otherwise display L copies of 'L' followed by X copies of 'X'

Step 15 If I = 4 and V = 0 display 'IV'
If I = 4 and V = 1 display 'IX'
Otherwise display V copies of 'V'
followed by I copies of 'I'

Exercise 3.2 Page 44

1. (i) 0.12 seconds
   (ii) 54 minutes
   (iii) n is too small for you to be able to tell what the run-time might be.
2. Probably $O(n^2)$
3. (i) $r = 41, q = 0; r = 38, q = 1; \ldots; r = 2, q = 13$
   (ii) Division by repeated subtraction
   (iii) $r = 4, q_1 = 0; r = 1, q_1 = 1; r = 11, q_2 = 0;$
   $r = 8, q_2 = 1; r = 5, q_2 = 2; r = 2, q_2 = 3$
   (iv) Second algorithm is a form of long division. It is more efficient but less transparent.
4. (i) 592
   (iii) A uses 30 592 additions, 0 subtractions and 10 000 comparisons.
   B uses 10 592 additions, 1110 subtractions and 10 000 comparisons.
   B is more efficient than A.
5. (i) HCF = 180
   (ii) One extra iteration in which A and B are swapped.
   (iii) 12 iterations, formula gives 12.0001 …

Discussion points Page 47

Optimal here means using the least number of bins. Depending on the context, other criteria may be more appropriate; for example it may be better to aim for an even balance of weights rather than having some full bins and some nearly empty bins, or other factors may need to be considered, such as the size of the boxes or (for food) their use by dates.

Using a complete enumeration will be very time consuming; a good solution that can be found quickly may be more useful than a best solution that cannot be found quickly.

Exercise 3.3 Page 48

1. Total on video = 63.2 which should fit onto four 16 GB USB sticks (16 × 4 = 64)
   First-fit decreasing algorithm gives:

Pro	P	J	L	Q	E	A	N	I	B
GB	8	5.6	4.8	4.8	4.4	4	4	3.6	3.2

Pro	K	M	O	D	C	H	G	R	F
GB	3	3	2.8	2.6	2.4	2.4	2	1.6	1

Bin 1   P(8) J(5.6) C(2.4)
Bin 2   L(4.8) Q(4.8) E(4.4) G(2)
Bin 3   A(4) N(4) I(3.6) B(3.2) F(1)

2. Sum of lengths is 82 m so at least 5 lanes are needed.
   First-fit decreasing, for example, gives
   $14 + 5; 12 + 8; 11 + 4 + 4; 10 + 4 + 4; 3 + 3$
   So 5 lanes are sufficient.
3. $3.5 + 2, 3.5 + 2, 3 + 2 + 1, 1.5 + 1.5 + 1.5 + 1.5, 1;$
   needs 5 pipes
4. (i) First fit gives: $2 + 1 + 6; 3 + 3; 5$ so 3 bags needed.
   (ii) First fit decreasing gives the optimal solution $6 + 3 + 1; 5 + 3 + 2$ so 2 bags needed.
5. (i) Session 1   F H H I A
       Session 2   B B B E E C D
       Session 3   C C G G G A A D D
   (ii) Session 1   F H I B E
        Session 2   H B E C
        Session 3   B C
   (iii) Each session must have A, B, C, D and G, leaving 36 minutes. But there are four 20-minute activities, so one session must have two of these. As there are only 36 minutes available there will not be a fit.
6. (i) Crate 1:   50 kg, 50 kg
       Crate 2:   50 kg, 40 kg
       Crate 3:   40 kg, 40 kg, 20 kg
       Crate 4:   30 kg, 30 kg, 30 kg
       Crate 5:   20 kg
       Five crates are used.
   (ii) Crate 1:   50 kg, 50 kg
        Crate 2:   50 kg, 30 kg, 20 kg
        Crate 3:   40 kg, 40 kg, 20 kg
        Crate 4:   40 kg, 30 kg, 30 kg

Exercise 3.4 Page 52

1. (i) 6 5 9 4 5 2
       5 6 4 5 2 9
       5 4 5 2 6 9
       4 5 2 5 6 9
       4 2 5 5 6 9
       2 4 5 5 6 9
   (ii) 6 5 9 4 5 2
        5 6 9 4 5 2
        4 5 6 9 5 2
        4 5 5 6 9 2
        2 4 5 5 6 9
2. (i) (a) 15
       (b) 21
       (c) $\frac{1}{2}n(n-1)$
   (ii) Because the dominant term is in n^2.

3 0.18 seconds

4 (i) $9 + 8 + 7 + 6 + 5 + 4 + 3 = 42$ comparisons

(ii) B A C E F G H I J D
92 81 76 82 45 51 93 71 62 43
B A E C G H I J F D
92 81 82 76 51 93 71 62 45 43
B E A C H I J G F D
92 82 81 76 93 71 62 51 45 43
$9 + 8 + 7 = 24$ comparisons

5 (i) $\{1, 2, 3\}$ 2 using shuttle and 3 using insertion;
$\{1, 3, 2\}$ 3 and 3; $\{2, 1, 3\}$ 2 and 3;
$\{2, 3, 1\}$ 3 and 2; $\{3, 1, 2\}$ 3 and 3;
$\{3, 2, 1\}$ 3 and 2.

(ii) (a) e.g. $\{7, 6, 5, 4, 3, 2, 1\}$ requiring
21 comparisons.
(b) e.g. $\{1, 2, 3, 4, 5, 6, 7\}$ requiring
21 comparisons.
(c) 45 comparisons

Discussion point Page 53

To adapt quick sort to sort the list into descending (decreasing) order, change 'less than or equal to' to 'greater than or equal to' (twice) and 'greater than' to 'less than' (twice) in Step 1, so Step 1 reads: The first value in the list is the pivot.

Excluding the pivot, pass along the list and write down each value that is greater than or equal to the pivot value, then write the pivot value and then write down the values that are less than the pivot.

At this stage the pivot is guaranteed to be in its correct position in the final list and can be marked in some way to indicate this. The pivot splits the list into two sublists: one containing the values that are greater than or equal to the pivot (excluding the pivot itself) and the other containing the values that are less than the pivot. It is possible that one of these sublists may be empty.

Activity 3.5 Page 56

Activity 3.6 Page 58

(a) 4, 5 or 6
(b) 4
(c) 4.83
(d) 4
(e) Yes, in that the worst case, 6 is greater than the average case, 4.83, in the same way that $4^2 > 4\log 4$. The size is too small to make further comments.

Activity 3.7 Page 58

(a) 6 (40 times), 7 (32 times), 8 (24 times),
9 (8 times) or 10 (16 times)
(b) 6
(c) 7.4
(d) 6 is the mode, 7 is the median
(e) Yes, in that the worst case, 10 is greater than the average case, in the same way that $5^2 > 5\log 5$. The size is too small to make further comments.

Exercise 3.5 Page 58

1 (i) 2 4 5 5 6 9

(ii) $6 + 3 = 9$

(iii) If the items are in reverse order, the pivot is compared with one less item each time, as one half of the pivot is empty. The number of comparisons is $\frac{1}{2}n(n + 1)$.

2 B, as it has $O(\log n)$ and that is less than the other two.

3 (i)

(ii) The space remaining is split into two pieces as shown in (i).

(iii) Sort by area and apply the algorithm in order of decreasing area.

4 Choose items 2, 3 and 5 with value 66.

5 (i) 13 comparisons

(ii) 21

(iii) 10 (e.g. 4625371)

6

Yes, as shown in the diagram. Ordering the boxes in decreasing order of volume, then placing them in the lowest, and right-most position, starting from the beginning each time until they are all placed. They would have to actually be packed in a different order.

7 (i) (a) 6
(b) 4
(ii) (a) 28
(b) 16

(iii) Doubling the size of the problem has a greater effect on the worst-case scenario than the more typical one. That is consistent with $O(n\log n) \subset O(n^2)$.

Chapter 4

Opening activity Page 62

For example, ABCDCBA: 14

Activity 4.1 Page 68

Assuming cubic complexity, $1.5 \times \left(\frac{20}{5}\right)^3 = 96$, or approximately a tenth of a second.

Exercise 4.1 Page 68

1   Bod–Tru 26; Tru–Pen 27; Bod–Bud 30;
     Bud–Oke 30; Oke–Exe 23; Exe–Tor 22;
     Bod–Ply 31 (or could choose any other of the
     31 weight arcs from Plymouth);
     Oke–Bar 31; Exe–Tau 35;
     Tau–Min 24

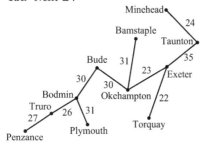

     Total weight = 279

2

	A	B	C	D	E	F	G	H
A	–	4	–	2	6	–	–	–
B	4	–	1	–	–	–	–	–
C	–	1	–	5	–	–	–	2
D	2	–	5	–	3	–	–	8
E	6	–	–	3	–	2	5	–
F	–	–	–	–	2	–	2	–
G	–	–	–	–	5	2	–	17
H	–	–	2	8	–	–	17	–

3
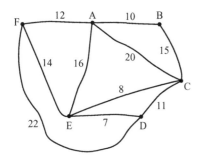

4   AB 10
     AF 12
     FE 14
     ED 7
     EC 8
     Total weight = 51

5   Approx. 32 seconds

6   ED 7
     EC 8
     AB 10
     CD 11 (exclude)
     AF 12
     FE 14

7   MST has length 82.

8   MST has length 66.

Discussion point Page 71

The shortest distances are from S, so that you are effectively being guided back to S. Were you to work forwards from S, there would be no way of knowing whether your route was going to be the shortest one to T.

Discussion point Page 71

We could reverse the process, and find the shortest route from T to one of $S_1, S_2, \ldots$

Activity 4.3 Page 72

Assuming quadratic complexity,

$2 \times \left(\frac{200}{20}\right)^2 = 200$ milliseconds, or approximately one-fifth of a second

Exercise 4.2 Page 72

1   (i)   SPQRT or SUVWT; 15
     (ii) SABFEDT; 8
     (iii) SBFJT; 12

2   (i)   L.A.–San Fran–Salt Lake–Omaha–Chicago; 42
     (ii) New Orleans–Chicago–Omaha–Denver; 34
     (iii) L.A–Santa Fe–Denver–Omaha–Chicago; 42
           (no better)
           New Orleans–El Paso–Santa Fe–Denver; 31
           (better)

3   C: 3 + 4 + 2 + 2 + 3 = 14

Activity 4.4 Page 74

A possible route is ABCBFBDFEDCA.
In order to ensure that the arcs CB and BF are repeated, a useful device is to repeat each one straightaway: thus CB comes after BC and FB comes after BF.

Activity 4.5 Page 75

A possible route is: ACABFEFDBCDEA
(which repeats AC and EF).

Activity 4.6 Page 75

A is an odd node, so you are happy to start there, if you don't need to finish there. B is an even node, so if you are to finish there then you need to turn it into an odd node by repeating a suitable path. Thus, the problem is the same as for category C(i), but with the four nodes to be dealt with being B, C, E and F (instead of A, C, E and F).

Activity 4.7 Page 75

The relevant pairings are

BC   4
BE   14 (BDE)
BF   7
CE   15 (CDE)
CF   11 (CBF)
EF   10

The possible ways of pairing up the nodes are

BC EF   4 + 10 = 14
BE CF   14 + 11 = 25
BF CE   7 + 15 = 22

The combination that gives the shortest total distance is thus BC EF, and the shortest route becomes 74 + 14 = 88.

Discussion point Page 75

You can select two of the odd nodes to be the start and end points, so that the path between them no longer has to be repeated. As AC is the shortest path between any pair of odd nodes, this is the one that you want to repeat, and so you choose E and F as the start and end nodes (or the other way round).

Activity 4.8 Page 78

$$(2n - 1)(2n - 3)... \times 5 \times 3 = \frac{(2n)!}{(2n)(2n - 2)... (6)(4)(2)}$$

$$= \frac{(2n)!}{2^n \times n(n - 1) ... (3)(2)(1)} = \frac{(2n)!}{n! \, 2^n}$$

When $2n = 20$, $\frac{(2n)!}{n! \, 2^n} = 654\,729\,075$

Exercise 4.3 Page 78

1 (i) e.g. 1, 3, 1, 2, 3, 4, 1 (length = 87)
   (ii) e.g. 1, 2, 3, 4, 1, 3 (length = 76)

2 For example, OYCYBLBOLCO, 1190 miles

3 For example, OYCYBLBOLCOL, 1247 miles

4 For example, YBOYCOLCLB, 979 miles

5 For example, Okehampton, Barnstaple, Bude, Okehampton, Bodmin, Truro, Plymouth, Exeter, Torquay, Plymouth, Bodmin, Bude, Barnstaple, Minehead, Taunton, Barnstaple, Exeter, Taunton, Exeter Okehampton; 655 miles

6 For example, Taunton, Exeter, Okehampton, Barnstaple, Bude, Okehampton, Bodmin, Truro, Plymouth, Exeter, Torquay, Plymouth, Bodmin, Bude, Barnstaple, Minehead, Taunton, Barnstaple, Exeter; 620 miles

7 (i) 83 + 18
   (ii) e.g. D $\to$ I $\to$ H $\to$ F $\to$ D $\to$ H $\to$ G $\to$ F $\to$ C $\to$ B $\to$ F $\to$ C $\to$ E $\to$ G $\to$ H $\to$ I $\to$ A $\to$ B $\to$ A $\to$ D
   (iii) 3 times

Discussion point Page 82

Prim's algorithm joins the nearest new node to **any** existing node, whereas the nearest neighbour algorithm joins it to the last node obtained.
Also, Prim's algorithm is designed to produce a tree and you don't return to the start node.

Activity 4.9 Page 83

Arcs are added in the following order.
Tau–Min 24
Min–Bar 38
Bar–Oke 31
Oke–Exe 23
Exe–Tor 22
Tor–Ply 31
Ply–Bod 31
Bod–Tru 26

Unfortunately, you are now stranded, and so the algorithm breaks down for this particular starting point. The algorithm is completed by taking each of the other nodes to be the starting point, in turn.

Exercise 4.4 Page 84

1 Remove A, 21; remove B, 21; remove C, 24; remove D, 22; remove E, 33; remove F, 21; remove G, 21; remove H, 24. The lower bound is 33.

2 Starting with A, 33; starting with B, can't return; starting with C, 33; starting with D, 33; starting with E, 33; starting with F, 33; starting with G, 33; starting with H, 33. The upper bound is 33.

3 Remove A, 66; remove B, 66; remove C, 62; remove D, 62; remove E, 63; remove F, 66. The lower bound is 66.

4 Starting with A, 74; starting with B, 69; starting with C, 74; starting with D, 74; starting with E, 69; starting with F, 74. The upper bound is 69.

5

	A	B	C	D	E	F
A	–	3	7	13	6	5
B	3	–	4	10	3	2
C	7	4	–	6	5	6
D	13	10	6	–	11	12
E	6	3	5	11	–	1
F	5	2	6	12	1	–

6 (i) Weston → Burnham → Bridgwater
→ Glastonbury → Wells → Bath → Cheddar
→ Weston; 95

(ii) Weston → Burnham → Bridgwater
→ Glastonbury → Wells → Cheddar → Bath
→ Weston; 103

7 (i) Birmingham → Gloucester → Hereford
→ Shrewsbury → Stoke → Sheffield
→ Nottingham → Northampton →
Birmingham; 357

(ii) The roads may be slower so the time is
longer than an alternative. It may not provide
appropriate stopping or refuelling points.

Exercise 4.5 Page 85

1 (i) 3

(ii) $5 \times 3 = 15$

(iii) $7 \times 5 \times 3 = 105$

(iv) 654729075

2 (i) B, D, F, G

(ii) Eulerian graphs have no odd nodes.

(iii) BD/FG = 550, BF/DG = 250,
BG/DF = 550
e.g. A → B → F → B → C → D → E → G
→ E → D → G → F → E → C → A
Length = 1950 + 250 = 2200

3 (i) Strawberry → orange → lemon → lime →
raspberry → strawberry; 73 minutes

(ii) No; this gives 77 minutes.

4 A_1 can be added to the table with distance to
and from A of ∞ and with distances to and
from B, C, D and E equal to the distances from
A to the towns. The solution to the travelling
salesperson problem will then be of the form A
★★★A1★★★A, and this can be interpreted as two
separate Hamilton cycles, one for each lorry.
The two tours are A → B → C → A and A →
D → E → A; total length 237

5 (i) (a) All of order 2

(b) All nodes are even so the network is
Eulerian.
Single rope needs only to be as long as the
sum of the four section lengths.

(ii) (a)

Node	A	B	C	D	E	F
Order	3	3	1	1	2	2

In traversing a network vertices are
encountered and left along different arcs.
So there needs to be an even number of arcs
incident upon each node.

(b) Some sections will have to have rope along
them twice, so the total length required will
exceed the sum of the lengths of the six
sections.

(c) Need to have AD and BC repeated, e.g.
C → B → F → A → D → A → E → B
→ C

(d) C → B → F → A → D → A → E →
B or C → B → E → A → D → A →
F → B or C → B → X → A → Y → B
→ Z → A → D, where {X, Y, Z} is any
permutation of {E, E, F} or {E, F, F}.
(There are six such possibilities.)

6 (i) A → B → D → E → C → F → A with
cost 640
Another tour is A → B → D → E → F → C
→ A with cost 537

(ii) $5! = 120$

(iii)

From \ To	A	B	C	D	E	F
A	–	65	80	78	110	165
B	75	–	97	55	113	130
C	80	90	–	70	90	340
D	90	65	90	–	75	250
E	110	90	80	45	–	82
F	165	130	320	195	100	–

(iv) e.g. B to E. Without taxes cheapest is BDE,
costing 60. With taxes cheapest is BE, costing
113.

(v) No difference. All taxes are incurred once and
only once, giving an increase of 140 on all
tours.

7 (i) The colour is equivalent to the town, the cost
of cleaning to the cost of travel, and using a
colour to visiting a town.

(ii) WYGBRW – 13
YGBRWY – 13
BGYRWB – 17 or BGYWRB – 13 or
BGRYWB – 14
GBRYWG – 12
RBGYWR – 13 or RGBYWR – 15

(iii) GBRYWG = WGBRYW

(iv) The two lowest weight arcs leading from the deleted node have to be an incoming arc and an outgoing arc, and this cannot be guaranteed in the case of a digraph.

Chapter 5

Opening activity Page 89

3 minutes

Activity 5.1 Page 90

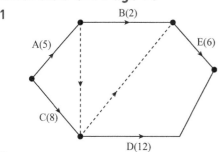

Exercise 5.1 Page 93

1

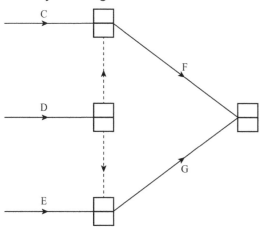

2

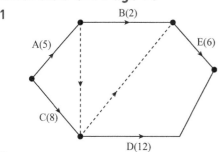

Minimum completion time is 5 days.

3

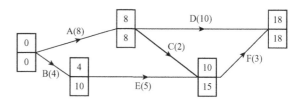

18 days

4 (i)

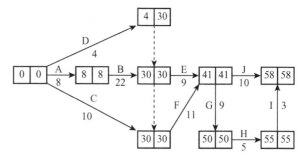

Minimum completion time is 58 weeks.

5 (i) (ii)

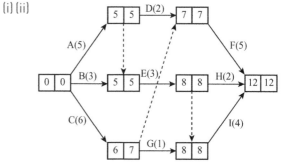

Minimum completion time is 12 days.

Activity 5.2 Page 95

17 hours

Exercise 5.2 Page 97

1 (i) Reduce A by 2 days.

(ii) Reduce A by 2 days, D by 5 days and F by 1 day.

2 A = 0, B = 1, C = 2, D = 0

3 (i)

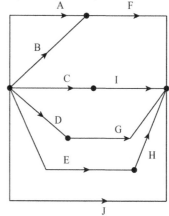

(ii) 45 minutes

(iii) C and I

Exercise 5.3 Page 100

1 (i)

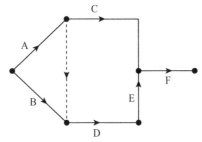

(ii) e.g.

Day	Harry	Nisha
1	A1	B1
2	D1	A2
3	E1	C1 B2
4	F1	D2
5	A3	B3 C2
6	D3	C3
7	E3	E2
8	F3	F2

2 (i) 44 days, A, B, C, G, I, L, N

(ii) D = 19 days (independent)
E = 20 days (independent)
F = 20 days (independent)
H and K = 1 day (interfering)
J and M = 2 days (interfering)

(iii)

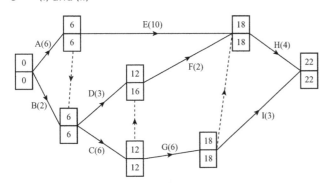

(iv) Delay J by 12 days to extend the project completion time to 54 days.

3 (i) and (ii)

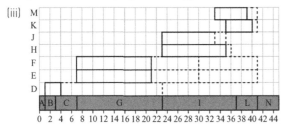

Minimum time to completion = 22 days
Critical activities are A, C, G and H.

(iii) Latest start time for D is 13 (16 − 3).
Earliest finish time for C is 12.
So the project can be completed in 22 days.

4 (i)

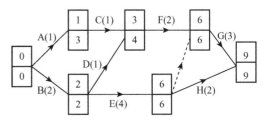

(ii) 9 days: B; E; G

(iii)

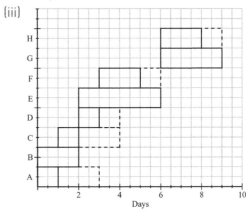

(iv) Start C at beginning of day 4.
Start F at beginning of day 5.

Chapter 6

Opening activity Page 103

Cost of materials, how much people are prepared to pay, number of potential customers, etc.

Activity 6.1 Page 106

At A, $3x + 4y = 24$ and $y = 3x \Rightarrow 15x = 24 \Rightarrow x = \frac{8}{5}$

so that A is (1.6, 4.8), with $P = 8$

Exercise 6.1 Page 108

1 $x = 3\frac{1}{7}, y = 2\frac{6}{7}, P = 8\frac{6}{7}$

2 $x = 4\frac{1}{4}, y = 3\frac{1}{6}, P = 55\frac{1}{12}$

3 Let x be the number of minutes spent walking.
Let y be the number of minutes spent running.
Maximise $D = 90x + 240y$
subject to $90x + 720y \leqslant 9000$
$x + y \leqslant 30$

Answer: $x = 20, y = 10, D = 4200$
Greatest distance is 4200 m = 4.2 km.

4 13 luxury and 17 standard, giving a profit of £335 000

5 $x = 1; y = 2; z = 3$ or $x = 0, y = 3, z = 3$

6   (i) $x + y \leqslant 4$

         $x \geqslant 2$

         $y \geqslant 0.6$

         $y \geqslant 0.25x$

   (ii)

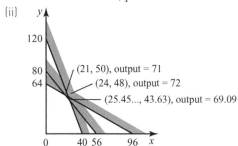

   (iii)   £3.2 million

   (iv)   2.3

   (v)   See heavy line on graph in part (ii).
Points of equal (fan) satisfaction

   (vi)   £3 382 000 in total; £618 000 less
£2 706 000 on the playing squad;
£206 000 more

7   (i)   Let x = amount of X produced (units) and
y = amount of Y produced (units).
Maximise $x + y$
subject to $15x + 5y \leqslant 600$

               $10x + 7y \leqslant 560$

               $8x + 12y \leqslant 768$

               $x \geqslant 0, y \geqslant 0$

   (ii)

       (21, 50), output = 71

       (24, 48), output = 72

       (25.45..., 43.63), output = 69.09

      21 units of X and 50 units of Y,
total output = 71 units

   (iii)   The second constraint becomes
$10x + 7y \leqslant 576$, making (24, 48) feasible.

   (iv)   Any further increase in availability of B is
irrelevant since the constraint is now not active.

8   Let x = amount of deep-mined (ores) and
y = amount of tonnes of opencast (tonnes).
Maximise $10x + 15y$ (or minimise $10x + 5y$)
subject to $x + y = 20\,000$

               $2x + y \leqslant 34\,000$ (chlorine)

               $3x + y \leqslant 40\,000$ (sulphur)

               $35x + 10y \leqslant 400\,000$ (ash)

               $5x + 12y \leqslant 200\,000$ (water)

Gives $x \approx 5700$; $y \approx 14\,300$, with water constraint
critical and the others redundant.

9   12 tables and 30 chairs; profit = £390
It is likely that the demand for chairs will be
greater than the supply.

10   (i)   Let x be the number of the cheaper jacket
and y be the number of the more expensive
jacket.
Maximise $P = 10x + 20y$
subject to $x + y \geqslant 200$

                $10x + 30y \leqslant 2700$

                $20x + 10y \leqslant 4000$

                $x \geqslant 0, y \geqslant 0$

   (ii) $x = 186$; $y = 28$; $P = 2420$

   (iii)   The purchase constraint exceeds the
maximum order allowed by the other
constraints (220 > 214).

Activity 6.2 Page 110

P will be maximised at the vertex C,
where $3x + y = 15$ and $x + 2y = 12$.

These equations can be solved to give C as (3.6, 4.2),
where $P = 19.8$.

Activity 6.3 Page 110

Neighbouring grid points give the following:

(3, 4): $x + 2y = 11$, $3x + y = 13$, $P = 18$

(3, 5): $x + 2y = 13$ (reject, as >12)

(4, 4): $x + 2y = 12$, $3x + y = 16$ (reject, as >15)

(4, 5): $x + 2y = 14$ (reject, as >12)

Exercise 6.2 Page 113

1   Node 1: UB = $8\frac{6}{7}$ $(3\frac{1}{7}, 2\frac{6}{7})$; LB = 7 (3, 2)

     Node 2 $(y \leqslant 2)$: UB = 8 (4, 2); new LB = 8 (4, 2)

     Solution: (4, 2); $P = 8$ (as 8 is effective UB)

2   Node 1: UB = $55\frac{1}{12}$ $(4\frac{1}{4}, 3\frac{1}{6})$; LB = 52 (4, 3)

     Node 2 $(x \leqslant 4)$: UB = $54\frac{2}{3}$ $(4, 3\frac{1}{3})$; LB = 52 (4, 3)

     Node 3 $(x \geqslant 5)$: UB = $40\frac{1}{3}$ $(5, \frac{2}{3})$ − reject, as < 52

     Node 4 $(x \leqslant 4 \ \& \ y \leqslant 3)$: UB = 52 (4, 3);

     LB = 52 (4, 3)

     Node 5 $(x \leqslant 4 \ \& \ y \geqslant 4)$: UB = 53 (3, 4);

     new LB = 53 (3, 4)

     Solution: (3, 4); $P = 53$

3   Node 1: LB = 49.5 (2.5, 6); UB = 51 (3, 6)

     Node 2 $(x \leqslant 2)$: LB = $56\frac{2}{5}$ $(2, 7\frac{1}{5})$ (reject, as > 51)

     Node 3 $(x \geqslant 3)$: LB = 51 (3, 6)

     Solution: (3, 6); $P = 51$

Exercise 6.3 Page 119

1

P	x	y	z	s_1	s_2	RHS
1	−9	−10	−6	0	0	0
0	2	3	4	1	0	3
0	6	6	2	0	1	8
1	$-\frac{7}{3}$	0	$\frac{22}{3}$	$\frac{10}{3}$	0	10
0	$\frac{2}{3}$	1	$\frac{4}{3}$	$\frac{1}{3}$	0	1
0	2	0	−6	−2	1	2
1	0	0	$\frac{1}{3}$	1	$\frac{7}{6}$	$\frac{37}{3}$
0	0	1	$\frac{10}{3}$	1	$-\frac{1}{3}$	$\frac{1}{3}$
0	1	0	−3	−1	$\frac{1}{2}$	1

Solution: $x = 1$, $y = \frac{1}{3}$, $z = 0$, $P = 12\frac{1}{3}$

2

P	w	x	y	z	s	t	RHS
1	−3	−2	0	0	0	0	0
0	1	1	1	1	1	0	150
0	2	1	3	4	0	1	200
1	0	−0.5	4.5	6	0	1.5	300
0	0	0.5	−0.5	−1	1	−0.5	50
0	1	0.5	1.5	2	0	0.5	100
1	0	0	4	5	1	1	350
0	0	1	−1	−2	2	−1	100
0	1	0	2	3	−1	1	50

Solution: $w = 50$, $x = 100$, $y = 0$, $z = 0$, $P = 350$

3

P	w	x	y	z	s	t	u	RHS
1	−3	−2	0	0	0	0	0	0
0	1	1	1	1	1	0	0	150
0	2	1	3	4	0	1	0	200
0	−1	1	0	0	0	0	1	0
1	0	−0.5	4.5	6	0	1.5	0	300
0	0	0.5	−0.5	−1	1	−0.5	0	50
0	1	0.5	1.5	2	0	0.5	0	100
0	0	1.5	1.5	2	0	0.5	1	100
1	0	0	5	$\frac{20}{3}$	0	$\frac{5}{3}$	$\frac{1}{3}$	$\frac{1000}{3}$
0	0	0	−1	$-\frac{5}{3}$	1	$-\frac{2}{3}$	$-\frac{1}{3}$	$\frac{50}{3}$
0	1	0	1	$\frac{4}{3}$	0	$\frac{1}{3}$	$-\frac{1}{3}$	$\frac{200}{3}$
0	0	1	1	$\frac{4}{3}$	0	$\frac{1}{3}$	$\frac{2}{3}$	$\frac{200}{3}$

Solution: $w = 66\frac{2}{3}$, $x = 66\frac{2}{3}$, $y = 0$, $z = 0$, $P = 333\frac{1}{3}$

4 (i) a = number of cuddly aardvarks made,
b = number of cuddly bears made,
c = number of cuddly cats made
First inequality models the furry material constraint.
Second inequality models the woolly material constraint.
Third inequality models the glass eyes constraint.
That would model a 'pairs of glass eyes' constraint.

(ii) The problem is an IP, so the number of eyes used will be integer anyway.

(iii) e.g.

P	a	b	c	s	t	u	RHS
1	−3	−5	−2	0	0	0	0
0	0.5	1	1	1	0	0	11
0	2	1.5	1	0	1	0	24
0	2	2	2	0	0	1	30
1	−0.5	0	3	5	0	0	55
0	0.5	1	1	1	0	0	11
0	1.25	0	−0.5	−1.5	1	0	7.5
0	1	0	0	−2	0	1	8
1	0	0	2.8	4.4	0.4	0	58
0	0	1	1.2	1.6	−0.4	0	8
0	1	0	−0.4	−1.2	0.8	0	6
0	0	0	0.4	−0.8	−0.8	1	2

Make 6 aardvarks and 8 bears giving £58 profit. 2 eyes are left over.

5 (i) Let x = number of mathematics books produced, y = number of novels produced and z = number of biographies produced.
Profit on each mathematics book is £10 − £4 = £6, on each novel is £3 and on each biography is £7.
Total profit (in £) is $6x + 3y + 7z$, this needs to be maximised.
Hence row 1: $P − 6x − 3y − 7z = 0$
Printing time (minutes) = $2x + 1.5y + 2.5z$, this must be $\leqslant 10\,000$
Hence row 2: $2x + 1.5y + 2.5z + s = 10\,000$
Packing time (minutes) = $x + 0.5y + 1.5z$, this must be $\leqslant 7\,500$
Hence row 3: $x + 0.5y + 1.5z + t = 7500$
Storage space (m³) = $300x + 200y + 400z$, this must be $\leqslant 2\,000\,000$
Hence row 4:
$300x + 200y + 400z + u = 2\,000\,000$

(ii)

P	x	y	z	s	t	u	RHS
1	−6	−3	−7	0	0	0	0
0	2	1.5	2.5	1	0	0	10000
0	1	0.5	1.5	0	1	0	7500
0	300	200	400	0	0	1	2000000
1	−0.4	1.2	0	2.8	0	0	28000
0	0.8	0.6	1	0.4	0	0	4000
0	−0.2	−0.4	0	−0.6	1	0	1500
0	−20	−40	0	−160	0	1	400000
1	0	1.5	0.5	3	0	0	30000
0	1	0.75	1.25	0.5	0	0	5000
0	0	−0.25	0.25	−0.5	1	0	2500
0	0	−25	25	−150	0	1	500000

Produce 5000 maths books at a profit of £30 000.

(2500 packing minutes spare and 0.5 m³ storage space spare)

(iii) £1.50 and 50p respectively

Exercise 6.4 Page 120

1 (i) a = amount of product A (tonnes),
b = amount of product B (tonnes),
c = amount of product C (tonnes)

Max $a + b + c$

st $3a + 2b + 5c \leqslant 60$
$5a + 6b + 2c \leqslant 50$
$a \geqslant 0, b \geqslant 0, c \geqslant 0$

(ii) e.g.

P	a	b	c	s	t	RHS
1	−1	−1	−1	0	0	0
0	3	2	5	1	0	60
0	5	6	2	0	1	50
1	−0.4	−0.6	0	0.2	0	12
0	0.6	0.4	1	0.2	0	12
0	3.8	5.2	0	−0.4	1	26
1	$\frac{1}{26}$	0	0	$\frac{4}{26}$	$\frac{3}{26}$	15
0	$\frac{8}{26}$	0	1	$\frac{6}{26}$	$-\frac{2}{26}$	10
0	$\frac{19}{26}$	1	0	$-\frac{2}{26}$	$\frac{5}{26}$	5

Make 5 tonnes of B and 10 tonnes of C.

2 (i) a = area of land (acres) used for crop A,
b = area of land (acres) used for crop B,
c = area of land (acres) used for crop C,
d = area of land (acres) used for crop D
$a + b \leqslant 20$ and $a + b + c + d = 40$ means that
$c + d \geqslant 20$

The coefficients of a, b, c and d in the objective are all positive so $a + b + c + d \leqslant 40$ will have no slack when the objective is maximised.

(ii)

P	a	b	c	d	s	t	RHS
1	−50	−40	−40	−30	0	0	0
0	1	1	0	0	1	0	20
0	1	1	1	1	0	1	40
1	0	10	−40	−30	50	0	1000
0	1	1	0	0	1	0	20
0	0	0	1	1	−1	1	20
1	0	10	0	10	10	40	1800
0	1	1	0	0	1	0	20
0	0	0	1	1	−1	1	20

20 acres to A and 20 acres to C, giving profit of £1800

3 (i) Let x = amount of product X (tonnes) and y = amount of product Y (tonnes).
Finance constraint: $400x + 200y \leqslant 2000$ or $2x + y \leqslant 10$
Staff constraint: $8x + 8y \leqslant 48$ or $x + y \leqslant 6$
Storage constraint: $x + 3y \leqslant 15$
$x \geqslant 0, y \geqslant 0$

(ii) $P = 320x + 240y$

(iii) $x = 4, y = 2, P = 1760$
Produce 4 tonnes of X and 2 tonnes of Y to give a profit of £1760.

(iv)

P	x	y	s	t	u	RHS
1	−320	−240	0	0	0	0
0	2	1	1	0	0	10
0	1	1	0	1	0	6
0	1	3	0	0	1	15
1	0	−80	160	0	0	1600
0	1	0.5	0.5	0	0	5
0	0	0.5	−0.5	1	0	1
0	0	2.5	−0.5	0	1	10
1	0	0	80	160	0	1760
0	1	0	1	−1	0	4
0	0	1	−1	2	0	2
0	0	0	2	−5	1	5

After 1st iteration: $x = 5, y = 0$
After 2nd iteration: $x = 4, y = 2$

4 Time spent travelling by lorry $= 3 - m - c$ (hours)

Distance travelled
$= 20m + 40c + 30(3 - m - c) = 90 - 10m + 10c$

Petrol used (gallons)
$= (20m \div 60) + (40c \div 40) + 30(3 - m - c) \div 20$

$= -\left(\frac{7}{6}\right)m - \left(\frac{1}{2}\right)c + \frac{9}{2}$

so $-\left(\frac{7}{6}\right)m - \left(\frac{1}{2}\right)c + \left(\frac{9}{2}\right) \leqslant \frac{5}{2} \Rightarrow 7m + 3c \geqslant 12$

Distance travelled by moped
$= 20m \Rightarrow 20m \leqslant 55 \Rightarrow 4m \leqslant 11$

Distance travelled by car and/or lorry
$= 90 - 30m + 10c \Rightarrow 6m - 2c \geqslant 7$

Maximise $P = -m + c$

Subject to $7m + 3c \geqslant 12$, $4m \leqslant 11$, $6m - 2c \geqslant 7$ and $m \geqslant 0$, $c \geqslant 0$

Solve graphically to get $m = 1.625$, $c = 1.375$

Do not use the lorry.

Put the moped in the car and travel for 1.375 hours (1 hour, 22.5 min), using $\frac{11}{8}$ gallons of petrol and travelling 55 miles. Then travel by moped for 1.625 hours (1 hour 37.5 min), using $\frac{13}{24}$ gallons of petrol and travelling 32.5 miles.

Distance travelled $= 87.5$ miles, all the time is used and $\frac{23}{12}$ gallons of petrol (leaving $\frac{7}{12}$ gallons unused).

5 (i) Let $x =$ number of xylophones, $y =$ number of yodellers and $z =$ number of zithers.

Max $\quad 180x + 90y + 110z$
st $\quad 2x + 5y + 3z \leqslant 30$
$\quad 4x + y + 2z \leqslant 24$
$\quad x \geqslant 0, y \geqslant 0, z \geqslant 0$

(ii)

P	x	y	z	s	t	RHS
1	−180	−90	−110	0	0	0
0	2	5	3	1	0	30
0	4	1	2	0	1	24
1	0	−45	−20	0	45	1080
0	0	4.5	2	1	−0.5	18
0	1	0.25	0.5	0	0.25	6
1	0	0	0	10	40	1260
0	0	1	$\frac{4}{9}$	$\frac{2}{9}$	$-\frac{1}{9}$	4
0	1	0	$\frac{7}{18}$	$-\frac{1}{18}$	$\frac{5}{18}$	5

The columns for x and y consist of 0's and a single 1, so x and y are basic variables. Reading down to the 1 and across to the final column gives $x = 5$ and $y = 4$. The column for z is not of this form, so z is non-basic and $z = 0$.

Make 5 xylophones and 4 yodellers.

(iii) Over two weeks ($x = 3$ and $z = 18$)

(iv) The feasible region is a convex polyhedron bounded by planes. The edges of the feasible region are lines where two planes meet. The optimal points found are the ends of one of these lines.

The line is on the objective plane $P = 1260$.

6 (i) Let w be the number of wardrobes made.
Let u be the number of drawer units made.
Let d be the number of desks made.

Max $\quad 80w + 65u + 50d$
st $\quad 5w + 3u + 2d \leqslant 200$
$\quad 4.5w + 5.2u + 3.8d \leqslant 200$
$\quad w + 0.75u + 0.5d \leqslant 50$
$\quad w \geqslant 0, u \geqslant 0, d \geqslant 0$

(ii)

P	w	u	d	s	t	u	RHS
1	−80	−65	−50	0	0	0	0
0	5	3	2	1	0	0	200
0	4.5	5.2	3.8	0	1	0	200
0	1	0.75	0.5	0	0	1	50
1	0	−17	−18	16	0	0	3200
0	1	0.6	0.4	0.2	0	0	40
0	0	2.5	2	−0.9	1	0	20
0	0	0.15	0.1	−0.2	0	1	10
1	0	5.5	0	7.9	9	0	3380
0	1	0.1	0	0.38	−0.2	0	36
0	0	1.25	1	−0.45	0.5	0	10
0	0	0.025	0	−0.155	−0.05	1	9

$w = 36$, $u = 0$, $d = 10$ and $P = 3380$ with $s = t = 0$ and $u = 9$ (not needed)

(iii) Make 36 wardrobes and 10 desks
Income $= £3380$
$9\,\text{m}^3$ of storage space spare

7 (i) Let $x =$ amount of product X (litres), $y =$ amount of product Y (litres) and $z =$ amount of product Z (litres).
Row 1 $\Leftrightarrow$ total profit $P = 10x + 10y + 20z$
(since $40\text{p} - 30\text{p} = 10\text{p}$, etc.)
Row 2 $\Leftrightarrow 5x + 2y + 10z \leqslant 10\,000$
(A's availability)
Row 3 $\Leftrightarrow 2x + 4y + 5z \leqslant 12\,000$
(B's availability)
Row 4 $\Leftrightarrow 8x + 3y + 5z \leqslant 8000$
(C's availability)

[ii]

P	x	y	z	s	t	u	RHS
1	−10	−10	−20	0	0	0	0
0	5	2	10	1	0	0	10 000
0	2	4	5	0	1	0	12 000
0	8	3	5	0	0	1	8 000
1	0	−6	0	2	0	0	20 000
0	0.5	0.2	1	0.1	0	0	1 000
0	−0.5	3	0	−0.5	1	0	7 000
0	5.5	2	0	−0.5	0	1	3 000
1	16.5	0	0	0.5	0	3	29 000
0	−0.05	0	1	0.15	0	−0.1	700
0	−8.75	0	0	0.25	1	−1.5	2 500
0	2.75	1	0	−0.25	0	0.5	1 500

Make 0 litres of x, 1500 litres of y and 700 litres of z giving a profit of 29000 pence = £290.

[iii] By 16.5p per litre

8 [i] 'He must paint the lower half of each wall in the more expensive paint.'

[ii] 'He has 350 m² of wall to paint.'
The constraint is covered by the requirement to maximise $P = x + y$.

[iii]

P	x	y	s	t	RHS
1	−1	−1	0	0	0
0	1.45	0.95	1	0	400
0	−1	1	0	1	0
1	−2	0	0	1	0
0	2.4	0	1	−0.95	400
0	−1	1	0	1	0
1	0	0	$\frac{5}{6}$	$\frac{5}{24}$	333.33
0	1	0	$\frac{5}{12}$	$-\frac{19}{48}$	166.67
0	0	1	$\frac{5}{12}$	$\frac{29}{48}$	166.67

166.67 m² using more expensive paint and 166.67 m² using less expensive paint. Coverage 333.33 m².

[iv] Coverage = 375 m² (187.5 of each type)

[v] The solution does not maximise the use of the more expensive paint.

Chapter 7

Discussion point Page 126

Player 1: worst outcome for A is 50; worst outcome for B is 30. Best of these is 50 so option A is the play-safe strategy for player 1.

Player 2: worst outcome for A is 40; worst outcome for B is 20. Best of these is 40 so option A is the play-safe strategy for player 2.

Activity 7.1 Page 126

The total for each cell is 100, so each player pays 50 to give the profit matrix $\begin{pmatrix} -25, 25 & 15, -15 \\ -10, 10 & 10, -10 \end{pmatrix}$ which is the zero−sum game $\begin{pmatrix} -25 & 15 \\ -10 & 10 \end{pmatrix}$.

Row maximin = max(−25, −10) = −10 and col minimax = min(−10, 15) = −10 so game is stable and has value −10. Row maximin = max(−25, −10) = −10 and col minimax = min(−10, 15) = −10 so game is stable and has value −10.

The value of the constant sum game is 40.

Exercise 7.1 Page 126

1 [i] $\begin{pmatrix} -2 & 1 \\ 0 & -3 \\ -1 & 4 \end{pmatrix}$

[ii] 4

[iii] 3

2 Play-safe strategy for player 1 is A and for player 2 is C.

[i] Play (A, C), player 1 wins 1 point and player 2 loses 1 point.

[ii] Play (A, A), each player wins 0 points.

[iii] Play (B, C), player 1 wins 2 points and player 2 loses 2 points.

3 Row maximin = max(−1, −3, −3, −5) = −1, col minimax = min(4, −1, 6) = −1.
Row maximin = col minimax so stable.
Player 1 chooses option A; player 2 chooses option B, value is −1.

4 [i] Player 1 chooses option B; player 2 chooses option B.

[ii] Player 1 chooses option A; player 2 chooses option C.

[iii] Not stable

[iv] Not stable

5 (i) Stable; value is 0
  (ii) Stable; value is 0
  (iii) Not stable
  (iv) Not stable

6 The maximum of the row minima is c.
  The minimum of the column maxima is also c.

Discussion point Page 132

This will mean that one of the lines is always below (or at the same level as) the other lines in the interval $0 \leqslant p \leqslant 1$. The pay-off matrix $\begin{pmatrix} 4 & 1 \\ 3 & 2 \end{pmatrix}$ gives the lines $3 + p$ and $2 - p$. The fact that $2 - p$ is always below $3 + p$ corresponds to the dominance of the 2nd column (it is always player 2's best option, whatever value is chosen for p). Choosing $p = 0$ ensures that player 1 always chooses the 2nd option (winning 2).

Exercise 7.2 Page 133

1 (i) B & C
  (ii) A & D

2 (i) e.g. 1st column dominated; then 1st row; reduced matrix: (0) [B,B]
  (ii) e.g. 2nd row dominated; then columns A, B and D are dominated; reduced matrix: (0) [A,C]
  (iii) No dominated strategies
  (iv) No dominated strategies

3 Row maximin = max(2, 1) = 2, strategy A is play-safe for player 1.
  Col minimax = min(3, 4) = 3, strategy A is play-safe for player 2.
  Row maximin ≠ col minimax so no stable solution.
  If player 1 chooses randomly between A and B, choosing A with probability p and B with probability $1 - p$:
  Expected pay-off when player 2 chooses A is $3p + (1 - p) = 1 + 2p$
  Expected pay-off when player 2 chooses B is $2p + 4(1 - p) = 4 - 2p$
  $1 + 2p = 4 - 2p \Rightarrow p = 0.75$
  Value of game = 2.5

4 Player 1 should not choose strategy C.
  This gives the reduced matrix $\begin{pmatrix} 1 & 2 \\ 3 & 0 \\ -1 & 4 \end{pmatrix}$
  Row maximin = max(1, 0, −1) = 1, play-safe for player 1 is strategy A.
  Col minimax = min(3, 4) = 3, play-safe for player 2 is strategy A.
  No stable solution.
  If player 2 chooses randomly between A and B, choosing A with probability q and B with probability $1 - q$:
  Expected pay-off when player 1 chooses A is $-\{q + 2(1 - q)\} = q - 2$
  Expected pay-off when player 1 chooses B is $-\{3q + 0(1 - q)\} = -3q$
  Expected pay-off when player 1 chooses B is $-\{-q + 4(1 - q)\} = 5q - 4$
  Optimum when $q = 0.5$
  Value of game = 1.5

5 (i) 3.4
  (ii) 4
  (iii) $3\frac{2}{3}$

6 (i) Row maximin = max(−1, 0, −1) = 0, play-safe for player 1 is strategy B.
    Col minimax = min(2, 2, 0, 4, 3) = 0, play-safe for player 2 is C
    Row maximin = col minimax so stable solution.
    Value of game = 0
  (ii) Row C dominates row A, column C dominates all the other columns; reduced matrix $\begin{pmatrix} 0 \\ -1 \end{pmatrix}$. In reduced matrix row A dominates row B; final reduced matrix is (0)

7 Row A dominates row D, row B dominates row E; $\begin{pmatrix} 2 & -1 & 0 & 4 \\ 0 & 6 & 2 & 2 \\ 4 & 0 & 1 & 1 \end{pmatrix}$
  In reduced matrix, column C dominates column D; $\begin{pmatrix} 2 & -1 & 0 \\ 0 & 6 & 2 \\ 4 & 0 & 1 \end{pmatrix}$
  Now row C dominates row A; $\begin{pmatrix} 0 & 6 & 2 \\ 4 & 0 & 1 \end{pmatrix}$
  In this reduced matrix,
  row maximin = max(0, 0) = 0
  and col minimax = min(4, 6, 2) = 2,
  row maximin ≠ col minimax so no stable solution.

8 $x \geqslant 1$

9 Player 1's strategy: choose A with prob. $\frac{1}{3}$

Player 2's strategy: choose A with prob. $\frac{5}{6}$

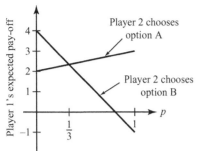

10 The value is 1.8.

11 Row C dominates row B, then column B dominates column A. Player 1 chooses randomly between A and C, choosing strategy A with probability 0.8; player 2 chooses randomly between B and C, choosing strategy B with probability 0.6

Value of game is 2.6

Exercise 7.3 Page 136

1 (i)

	A	**B**
A	⑨, ⑧	5, 7
B	7, 6	⑧, ⑩

Nash equilibrium positions: A, A & B, B

(ii)

	A	**B**
A	5, ⑦	③, ⑦
B	⑥, 4	1, ⑥

No (strict) Nash equilibrium positions

(iii)

	A	**B**
A	⑩, ⑧	7, 7
B	6, ⑦	⑧, 5

Nash equilibrium position: A, A

(iv)

	A	**B**
A	6, ⑨	④, 6
B	⑧, 6	2, ⑦

No Nash equilibrium positions

2

	A	B	row minima
A	9	1	1
B	10	2	②
col. maxima	10	②	

B, B is a stable solution. If player 1 has chosen B, then player 2 will not want to change to A (because 2 is the row minimum), and similarly if player 2 has chosen B, player 1 will not want to change to A (because 2 is the column maximum). Hence, B, B is a Nash equilibrium.

3 The Nash equilibria are circled.

	A	**B**
A	②	3
B	1	2
C	②	4

Approach 1: Row C dominates row A weakly, and so the matrix can be reduced to:

	A	**B**
B	1	2
C	2	4

Then column A dominates column B, so that the matrix can be further reduced to:

	A
B	1
C	2

And then row C dominates row B, so that the outcome is C, A.

Thus the Nash equilibrium A, A has been lost.

Approach 2: Column A dominates column B, and so the matrix can be reduced to:

	A
A	2
B	1
C	2

As row B is dominated by rows A and C, the matrix can be further reduced to:

	A
A	2
C	2

This gives the two Nash equilibria as possible outcomes.

Exercise 7.4 Page 140

1 Maximise $P = v - 1$, subject to

$v \leqslant 2p_1 + 3p_2 + 5p_3$,

$v \leqslant 4p_1 + p_2$,

$p_1 + p_2 + p_3 \leqslant 1$

and $p_1, p_2, p_3 \geqslant 0, v \geqslant 0$

2 The value is 2.5.

3 (i) 3.4

(ii) 4

(iii) $3\frac{2}{3}$

4 Maximise $P = v - 3$, subject to the constraints

$v \leqslant p_1 + 4p_3$,

$v \leqslant 2p_1 + 3p_2 + p_3$,

$p_1 + p_2 + p_3 \leqslant 1$

$(p_1, p_2, p_3 \geqslant 0, v > 0)$.

simplex equations:

$P - v = -3$

$v - p_1 - 4p_3 + s = 0$

$v - 2p_1 - 3p_2 - p_3 + t = 0$

$p_1 + p_2 + p_3 + u = 1$

5

P	v	p	q	s_1	s_2	s_3	RHS	
1	−1	0	0	0	0	0	−3	①
0	1	−7	−2	1	0	0	0	②
0	1	0	−5	0	1	0	0	③
0	0	1	1	0	0	1	1	④

6 $x \geqslant 2$

7 (i) Janet should choose 30 and John should choose 24. Janet wins 6.

(ii) 17